Justice Downwind

JUSTICE DOWNWIND,

America's Atomic
Testing Program
in the 1950s

HOWARD BALL

New York   Oxford
OXFORD UNIVERSITY PRESS
1986

Oxford University Press

Oxford   New York   Toronto
Delhi   Bombay   Calcutta   Madras   Karachi
Petaling Jaya   Singapore   Hong Kong   Tokyo
Nairobi   Dar es Salaam   Cape Town
Melbourne   Auckland

and associated companies in
Beirut   Berlin   Ibadan   Nicosia

Published by Oxford University Press, Inc.,
200 Madison Avenue, New York, New York 10016

Library of Congress Cataloging in Publication Data
Ball, Howard, 1937–
Justice downwind.
Includes index.
1. Nuclear weapons—Testing.
2. Radioactive fallout—Utah—Physiological effect.
3. Radioactive fallout—Nevada—Physiological effect.
4. Liability for nuclear damages—Utah.
5. Liability for nuclear damages—Nevada.   I. Title.
U264.B35   1985   363.1'79   85-8777
ISBN 0-19-503672-7

Printing (last digit):   9 8 7 6 5 4 3 2 1

Printed in the United States of America
on acid free paper

For Dorothy Rabin

Preface

This is a story about atoms, about the making and the testing of nuclear weapons, about the dangers and the uncertainties, and the contradictions that abound when human beings create such awesome, once-unimaginable power. It is about the dramatic tensions and conflicts facing medical researchers as they try to ascertain the consequences of the atomic testing program at the southeastern Nevada test range. It is also a story about governmental responsibility and possible irresponsibility, about legal relationships and obligations between the federal government and its sovereign citizens, and about social values and public policymaking in a democracy. This is a story about the new technology of nuclear power, its varied impact on our social system, and the dark side of that new technology.

Justice Downwind is also about physicists and other nuclear scientists, many of them refugees from Nazi Europe; farmers, sheepherders, small-town schoolteachers and their schoolchildren; lawyers—government and private counsel for injured plaintiffs; federal judges, doctors, medical researchers, federal bureaucrats, congressional staffers, and their bosses—the politicians. It is a story with many locations, from the small desert towns in the Southwest—Kanab, Bunkerville, St. George, and Cedar City—and the "science towns" of Hanford, Los Alamos, Oak Ridge, and Brookhaven, to the large cities on both coasts—Los Angeles, New York and, especially, Washington, D.C.

Above all, the story of America's atomic testing program during the 1950s is a story full of conflicts, some of which were tragic. Also present in the atomic testing story, from the initial development of the atomic bomb at Los Alamos, New Mexico during the early 1940s, to the contemporary political, legal, medical, and ethical controversies over the impact of low levels of ionizing radiation on humans, is the emergence of many controversies, many of which were never resolved.

The scientists' horror at the potential destructiveness of their bomb, yet their continued work on its development and improvement, was one such dilemma. Another was the military's assertion that the atomic secret would remain a secret, while all the time it was not a secret to be kept. While the Atomic Energy Commission was responsible for ensuring proper safety standards so that the people who lived downwind of the atomic testing would not be endangered, the AEC always rushed efforts to develop a nuclear superweapon that would maintain America's atomic supremacy.

And the downwinders' loyalty to their government and faith in its representatives who were performing the tests was occasionally questioned by citizens who voiced great concern about the tests and who wished these events would take place somewhere else.

Medical experts clashed with each other in their efforts to explain the impact of the atomic testing on the people who lived downwind; the law of tort liability was confusing on the question of governmental immunity in such instances. These dilemmas and conflicts found their way into the political arena and, for almost a decade, legislators have debated the critical issue of compensation for persons who have been injured by low-level radiation exposure from the above-ground atomic blasts.

Justice Downwind examines these various conflicts—medical, scientific, legal, political, and normative—in an effort to explain what happened as America entered the Atomic Age in the 1940s. A complex contradictory picture emerges from a review of the documents and newspaper accounts of the era, replete with spies, suspicion, and secrecy. It involves complex relationships between nuclear scientists, military weapons experts (two natural antagonists who were primarily responsible for developing the atomic weapons systems deployed in the 1950s), civilian bureaucrats, political appointees, regional and national politicians, medical researchers, local doctors, and the inhabitants of the regions themselves.

Justice Downwind also addresses the fundamental questions of wrongful injury and remedies that have been raised since the development of the atom bomb. This drama involves, primarily, the two sets of people directly involved in the above-ground testing program at the Nevada Test Site: (1) the local AEC management personnel (and the scientists, technicians, and military liaisons, who supported and advised them) responsible for scheduling the many test shots that occurred in the Nevada desert during the 1950s, and (2) the local residents, over 100,000 persons, who lived downwind of the test site in the small towns dotting Nevada, Utah, and Arizona.

Years later, in 1978, these downwind citizens alleged that the government's above-ground testing of atomic weapons had led to illness and death. Specifically, they claimed that the managers of the above-ground testing program at the Nevada Test Site had been negligent in their efforts to provide the downwinders with safety information that would have minimized the hazards of radiation.

In our justice system, there are a number of remedies available to persons who claim injuries from negligent federal governmental action. One is the administrative remedy. It is used by claimants who seek compensation from the agency whose negligent action, they allege, is responsible for their injuries and the wrongful deaths of the family members. A second remedy available for injured persons is the adjudicative route in the federal district courts. The Federal Tort Claims Act, passed in 1946, allows persons to file civil suits against the United States government alleging that (1) governmental agents acted negligently and (2) that this negligence caused the injuries and the deaths alleged by the plaintiffs. A third basic remedy is the political one. For example, in 1969, the U.S. Congress passed compensatory legislation for coal miners afflicted with "black lung" that provided aid for persons who had been adversely affected by specified environmental dangers alleged to have been caused by governmental negligence.

Present in the legal, political, and moral battle between U.S. citizens and the U.S. government is a fundamental principle of American law which also lies at the heart of the three remedies mentioned above: *"The law imposes a duty on everyone to avoid acts in their nature dangerous to the lives of others."* [*Thomas v. Winchester*, 6 NY 397 (1852)]. To the degree that this principle is at work in this raging controversy, *Justice Downwind* is also about the flexibility of American justice. If people are wronged by agents of the federal government, then, somehow, a remedy must be devised that would restore justice in our society.

Fundamentally, then, *Justice Downwind* is a story of how a democratic society successfully beat a totalitarian society in the Herculean effort to create atomic weapons; how that democratic society managed to create the new world of military atomic power; and how that same democratic society has had to continuously struggle with the medical, political, and ethical consequences of this terrible new technology.

Salt Lake City   H.B.
June 1985

Acknowledgments

Many people have helped me as I have explored the complex worlds of nuclear physics, microbiology, biochemistry, medical epidemiology, cancer research, psychology, law, administration and politics. I want to take this opportunity to thank all of them as they have provided the insights necessary for my understanding in these areas, enabling me to address the many dynamic aspects of the Nevada Test Site story.

The University of Utah medical community was an important starting point for me in my efforts to understand the associations between radiation exposure and subsequent illnesses. Dr. Lynn Lyon of Family and Community Medicine was extremely helpful; my interviews with him and his research associate, Dr. Ken Smith, have proven to be invaluable. The Utah Cancer Registry staff at the University of Utah provided me with vital statistical information that was crucial to my work. Dr. McDonald Wrenn, Radiobiology; Dr. Tip Taylor, Nuclear Medicine; and Dr. Robert Fineman, Pediatrics, were very helpful, as was Dr. Chase Peterson, President, University of Utah (formerly Vice President for Health Sciences). Dr. Naomi Alazraki, Nuclear Medicine, Veterans Administration Hospital, Salt Lake City, was also very helpful to me, as were Robert and Gail Weinstein, Stony Brook, New York.

Dr. Jay Olshansky, Department of Sociology at the University of Utah, provided me with important insights into the world of medical epidemiology, while psychology professors Fred Rhodewalt and Irwin Altman, University of Utah (Dr. Altman is also Vice President for Academic Affairs), and Dr. Arthur Baum, Uniformed Services University of the Health Sciences in Bethesda, Maryland, furnished me with important ideas and views of the possible psychological impact of the radiation fallout on the citizens who lived downwind of the Nevada Test Site.

I valued the opportunity to speak with Henry Gill, government attorney with the Department of Energy, who prepared and argued the *Allen* case in federal court. The Chief of Legal Staff in the Department of Justice on nuclear radiation issues, Ralph Johnson—then Assistant U.S. Attorney of Utah—also gave me invaluable insights into the government's approach to the nuclear testing litigation. Also, Assistant Director Paul F. Figley, Torts Branch, U.S. Department of Justice, assisted me in locating some recent U.S. District Court opinions. I appreciated his help very much.

I also want to acknowledge the immense help I received from Russell

Kearl, former clerk for U.S. district court judge Bruce S. Jenkins, and his administrative secretary, Merrie Lynch, when I reviewed the mounds of documents, including seventeen volumes of trial transcript, in the judicial chambers. Judge Jenkins offered me generous assistance in developing the litigation chapters, and through these contacts I have come to greatly appreciate and respect his outstanding legal scholarship as a federal district court judge. Students of federal judicial decision-making at the trial level would be extremely well served by reviewing the opinions of this fine human being. Local attorneys for the plaintiffs, including Sandor Dolowitz, also helped me in my endeavors, and I appreciate their assistance very much.

Personal interviews with former Governor Scott Matheson (D-Utah) provided me with important insights into the politics of opening up the then confidential and hidden Atomic Energy Commission and Public Health Service files. These thousands of pages of documents revealed to the American public a pattern of bureaucratic activity that has been labeled by one federal district court judge, Sherman Christensen, as deceptive and fraudulent. Governor Matheson's position on this matter has been clear and forthright: the government made grievous mistakes and committed actions that are subject to civil liability damages—for which it should pay in full. He has been a champion of human rights since the facts began to emerge in 1978 and is an exemplary citizen-politician.

I want to especially thank Janet C. Gordon, Director of Citizens Call and her sister, Mary Lou Melling, for allowing me to interview them about the Nevada Test Site's impact on the small communities in Nevada, Utah, and Arizona. These women have been principal fighters for health care for the people downwind of the site. Both have experienced personal anguish and have transcended these tragedies to work hard, without pay, on behalf of others in the downwind communities.

In addition, I appreciated the opportunity to meet with Jay Truman, presently of Salt Lake City, Utah, to discuss the nuclear testing dilemma. Truman is president of the Downwinders, another small citizens' group seeking an end to nuclear testing and for compensation for its victims.

I have had a number of very fine research associates on this project. Kathanne Greene, a Ph.D. candidate in Political Science at the University of Utah, has been extremely valuable in the restoration of the recent history of the events surrounding the above-ground testing program at the Nevada Test Site. In addition, her interviews with Jay Truman, Bennie Levy, and Janet Gordon were of great value to me in putting together the chapter on the people who lived downwind of the testing activities.

Kenneth Verdoia, a Congressional Fellow of the American Political Science Association, 1983–1984, is an outstanding television analyst for the PBS television station KUED, in Salt Lake City, Utah. While in Washington, D.C., Ken provided me with important legislative documentation and had an excellent interview with one of the chief staffers for the Labor and Human Resources Committee of the U.S. Senate. He, too, is a graduate student at the University of Utah. Having been their instructor in a graduate Constitutional Law seminar, I can attest that Kathanne and Ken are outstanding students as well as fine investigative researchers. I am indebted to them and take this opportunity to tell them how much I have valued their help.

Three other fine graduate students in the Political Science Department at the University of Utah, Melanie Cherry, Cheryl Zimmerman, and Jane Simpson, assisted me in reviewing and reconstructing the events of the 1940s and 50s in southern Utah. They were of immense help to me in this task.

I also wish to acknowledge my students in the Fall, 1983 Judicial Process class who focused on the *Allen* litigation for their research projects. In particular, the outstanding work and provocative questions of Laverne Snow, Bonnie Barker, Brian Nelson, Gary Meyers, Cornell Clayton, Robert Norton, and Randy Snow prodded me into thinking anew about certain ideas. I enjoyed working with them and thank them once again for their constructive thoughts and probing questions. Sheryl Ball was an early reviewer and critic of the book. I greatly appreciated her help.

I thank Tom Patterson, cartographer for the Department of Geography at the University of Utah, for the quality maps he provided for the book.

Academic colleagues who share my interests in judicial policy-making were also extremely helpful. They include Dean Mann, University of California, Santa Barbara, and especially Phil Cooper, State University of New York, Albany. Bob Benedict, Political Science, at the University of Utah, was very helpful on legislative politics and deficit budgets. I thank them and the many others who have interacted with me at professional meetings where these issues were discussed and debated.

I want to thank my family, my wife Carol and my three beautiful daughters, Melissa, Sheryl, and Sue, for sharing these intensely dramatic events with me during two past very hectic years. Putting a book together is typically a stressful time; I compiled many of the materials while assuming the new and challenging responsibilities of Dean of the College of Social and Behavioral Science at the University of Utah. It was a time frame I would not want to go through again! My children and my wife were really terrific and I want them to know how much their support and forbearance meant to me.

Finally, I want to pay homage and fealty to two terrific editors at the University of Utah and at Oxford University Press. Jill Maryon, a marvelous editor and fine secretary in my office, has been invaluable to me in this project. As a long-time Utah resident, she has a feel for the people of the area, and on a number of occasions has conveyed her insights to me. She is also an excellent typist, capable of following my arrows and meanderings on both sides of draft sheets with great aplomb. I just don't know what I would have done without her eager assistance. I want to take this opportunity to thank her for all her efforts.

My other editor, at Oxford University Press, New York, is Susan Rabiner. I appreciate her insights on the context and texture of the book. I owe her a great debt of gratitude for gracefully providing me with some very interesting new and rewarding directions. I also appreciate the very valuable editorial efforts of Susan Meigs and Rachel Toor. It has been an enlightening and enjoyable experience for me to have worked with such competent, consummate, and friendly people at Oxford University Press.

I want to thank Margaret Sullivan, my Administrative Assistant, for her continuing assistance. Finally, a big thank you to Rosalie Cline, my secretary. Rosalie has been of immense help to me as we prepared the final drafts and proofed the galleys and page proofs. I greatly value her editing skills and her enthusiasm.

I know that there are others that have interacted with me as I have relived the above-ground testing activities of the Nevada Test Site and their impact on the downwind population, for the past three years. I am grateful for their help and want to indicate publicly that if any errors of interpretation have occurred they are mine alone. I have used my judgment in making the evaluations and in drawing these conclusions. I believe the documents and other evidence presented in the book buttress my positions on these matters. I will let the reader be the final judge.

Contents

Justice Downwind

1 "My God, It Worked!": The Development of the Atomic Bomb, 1939–1946

1939

In 1939, America was approaching the end of the Great Depression; John Steinbeck's portrayal of a dislocated America, *Grapes of Wrath*, was a national best-seller, the blockbuster film "Gone With The Wind," depicting an earlier period of turmoil in America's history, opened to packed theatres from Atlanta to California; a youthful Kate Smith began singing Irving Berlin's patriotic "God Bless America;" Mary Martin stormed Broadway; and a skinny Italian-American singer, Frank Sinatra, joined the Harry James band.

Then, in September, World War II began with Nazi Germany's attack on Poland. The Soviet Union marched into Poland from the east, and, within a few weeks, the Russians and the Germans (former mortal enemies) had devoured and divided Poland between them. President Franklin D. Roosevelt immediately declared America's "neutrality" in the European struggle but began to build up America's military forces.

World War II would prove to be not only a conflict of military against military, but of scientist against scientist. For thousands of years, the atom, of which all matter is comprised, had been the smallest particle known to science. But in the early 1920s, scientists discovered even smaller particles existing in each atom: protons, electrons and neutrons. The scientists speculated about the possibilities and consequences of splitting certain types of atoms. Nobel Prize winner Albert Einstein was one of a number of physicists who hypothesized that mass could be converted to highly unstable energy through the phenomenon of fissioning—splitting certain types of atoms, such as those of the element uranium. Although only 1/100,000 the volume of the atom, the nucleus contains most of the atom's mass—the protons and the neutrons. Physicists theorized that fission could be accomplished by bombarding the nucleus with a beam of neutrons shot at high velocity. This bombardment would split the uranium's nucleus, releasing some of its neutrons, which in turn would bombard other nuclei.[1]

Unlike the power generated by sunlight or water, or power from chemical reactions (coal and oil),[2] the release of "atomic" power was dependent on the initial chain reaction, whereby atoms were continuously and rapidly split.[3] Once a nucleus absorbs a neutron (in a millionth of a second) it splits and releases additional neutrons indefinitely. The reaction is self-sustaining when the number of neutrons released in a given time equals or exceeds the number of neutrons lost by absorption in non-fissioning material or by escape from the system. (See Figure 1.)

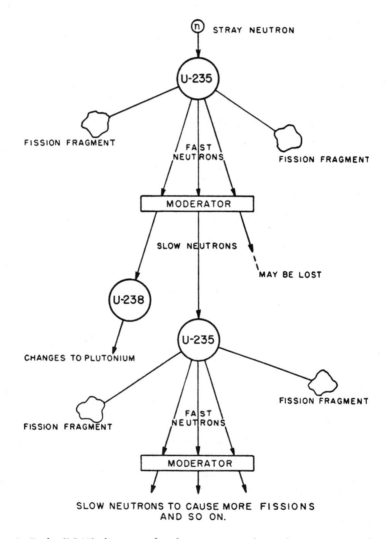

Figure 1. Early (1945) diagram of a chain reaction drawn by a scientist who would later serve as a commissioner of the Atomic Energy Commission, Henry D. Smyth. (Source: Smyth, *Atomic Energy for Military Purposes*, p. 35.)

During the 1930s, scientists invented new machines, such as cyclotrons, generators, and accelerators, that freed neutrons and propelled them at a velocity sufficient to set up this chain reaction. In 1939, Ernest Lawrence of the University of California, Berkeley received the Nobel Prize in Physics for inventing the cyclotron. Soon he would also play a major role in the development of the atomic bomb. At Princeton that same year, Danish physicist Niels Bohr calculated that "only a rare form of uranium, called U-235, making up about one percent of the elements in nature, could sustain a chain reaction."[4]

Before the splitting of the atom in 1938, it was difficult for scientists to conceive of such a transformation, because they had always believed elements could be altered only in form, not transformed into other elements. But now the research of these twentieth-century alchemists was directed toward converting matter (mass), through the process of fission, into fantastic amounts of energy. This harnessing of atomic power opened up new vistas along two divergent paths: peaceful purposes and the use of super-weapons of war.[5]

In December, 1938, two German physicists working at the Kaiser Wilhelm Institute in Berlin actually split a uranium atom and transformed it into barium. In the "remnants of bombarded uranium, they found barium—an element that weighs about half as much as uranium. There had been no barium in their sample to begin with. Apparently, some uranium nuclei had been cut in two."[6] Otto Hahn and Fritz Strassmann had succeeded in transforming matter. Their pre-World War II discovery of nuclear fission "set in motion the forces of science and politics that culminated seven years later in Hiroshima and Nagasaki."[7]

Their accomplishment was tremendous news and they spread the word to their friends in the German scientific community. Hahn also immediately revealed the results of his fission experiments in a letter to Lise Meitner, an Austrian-Jewish physicist who had fled from Germany to Scandanavia. He had worked for 30 years with Meitner, "who was a respected and much beloved physicist at the Kaiser Wilhelm Institute."[8] Meitner's nephew, Jewish physicist Otto Frisch, happened to visit her at this time and found her "brooding over a letter from Hahn." He recalled walking up and down in the snow with Lise, puzzling over how a single neutron could split in half a very large uranium nucleus.[9] They concluded that Danish physicist Niels Bohr, who had performed ground-breaking work on the atom's structure and received a Nobel Prize in 1922, might have the answer. Meitner, fearful that Nazi Germany might immediately begin harnessing atomic power, sent Frisch to Copenhagen to discuss the breakthrough with Bohr (who was later to flee Nazi-occupied Denmark). Frisch managed to meet with Bohr just before Bohr's departure for Princeton, where he planned to do research, so Bohr arrived in America with news of the dramatic developments taking place in Germany. He spoke mostly with European emigré theoretical physicists, such as Leo Szilard and Enrico Fermi (both involved in research at Columbia University), and Eugene P. Wigner and Albert Einstein (at Princeton's Institute for Advanced Study) about the fissioning experiments of Hahn and Strassmann. Bohr also discussed the dreaded consequences of the Germans developing atomic power before the free world did.[10]

Thus, two ironies that would fundamentally redirect the course of human history made themselves felt. The first irony was the fact that it was mainly Jewish physicists who transmitted the news to the West about the nuclear fission experiments in Germany.[11] The second irony lay in the fact that the refugee scientists in this country were restricted from military research because of their alien status, and were engaged in theoretical research only on atomic power. So the Fermis, Wigners, Tellers, Weisskopfs, Szilards, and Einsteins, blocked from radar research, which had a more demonstrably military application, were given ample time to speculate on the development of atomic power for wartime use against Hitler.[12] On such twists of fate was the future course of the war determined.

In 1939, most American nuclear physicists "had not yet become accustomed to thinking of new scientific truths in terms of their military applications," confessed General Leslie Groves, military leader of the Manhattan Project, in his memoirs.[13] However, because the foreign-born physicists had all experienced the horrors of totalitarianism and had fled Nazism, they—more than their American counterparts—took the lead in the early efforts to restrict scientists from publishing research about nuclear physics. These brilliant European refugees were committed to developing a super weapon before Hitler's nuclear physicists did.[14] Scientific knowledge was no longer to be shared worldwide.

On August 2, 1939, Albert Einstein took the commitment of the refugee scientists one step further. In a letter to President Roosevelt, Einstein emphasized the seriousness of the problem posed by Germany's gaining the upper hand in atomic power. Einstein's letter (drafted by Leo Szilard) was delivered to Roosevelt in October, 1939 (one month after World War II had begun in Europe) by Albert Sachs, a friend of Roosevelt.

Einstein's message to Roosevelt clearly pointed out that:

> The element uranium may be turned into a new and important source of energy in the immediate future. . . . It may become possible to set up a nuclear chain reaction in a large mass of uranium, by which vast amounts of power and large quantities of new radium-like elements would be generated. Now it appears almost certain that this could be achieved in the immediate future. This new phenomenon would also lead to the construction of bombs, and it is conceivable—though much less certain—that extrememly powerful bombs of a new type may thus be constructed. A single bomb of this type, carried by boat and exploded in a port, might very well destroy the whole port, together with some

of the surrounding territory. . . . In view of this situation, you may think it desirable to have some permanent contact maintained between the Administration and the group of physicists working on chain reactions in America. . . . I understand that Germany has actually stopped the sale of uranium from the Czechoslovakian mines which she has taken over. That she should have taken such early action might perhaps be understood on the ground that [the] Kaiser-Wilhelm-Institute in Berlin [is repeating] some of the American work on uranium.[15]

Einstein, Szilard, and Sachs wanted the President to commit money and support to their effort both to secure adequate supplies of uranium and to speed up the nuclear discovery process by enlarging the scope of individual and industrial laboratory research. Roosevelt responded immediately to Sachs in a meeting Sachs attended with Szilard: "Alex, what you are after is to see that the Nazis don't blow us up."[16]

Roosevelt, recognizing the urgency of the issue, acted quickly and created an Advisory Committee on Uranium to investigate the issues raised by Einstein and Sachs. There had been little formal contact between the government and the scientific community before 1939. But now began the fateful government science interaction that would lead to a final irony: atomic sunbursts would be detonated not over the cities of Hitler's Germany, but first, above the Japanese cities of Hiroshima and Nagasaki and, when the winds were blowing in an easterly direction, close to the American cities downwind of the Nevada Test Site.

Developing the Bomb

"Monstrous obstacles"[17] confronted the scientists and military men assigned the task of beating the Nazis to the secret of developing the atomic bomb. The United States had to gear up for the almost impossible research and development tasks associated with the military use of atomic energy. For nuclear physicists to move from the purely theoretical speculation about splitting an atom "to an operating plant of commercial size was a phenomenal achievement, an even greater venture into the unknown than the first voyage of Columbus."[18] To make it all happen, three major tasks had to be completed in reasonable synchrony—before the Nazi physicists accomplished the same: (1) a controlled chain-reaction had to be achieved to prove their theoretical assertions about atomic power; (2) a sufficient amount of raw materials (uranium and plutonium) had to be developed and manufactured to "power" the weapon; and (3) the weapon itself—a workable atomic bomb—had to be designed and built. While the scientists hypothesized about the "enormously intense radiations"[19] that a successful chain reaction in a bomb casing would produce, accidental exposure to radioactive material was not seen as a major problem. And, in fact, by using prudent safety measures, there were very few radiation accidents. There was also a very peculiar perception about this health danger. General Leslie R. Groves, in charge of the Manhattan Project, stated succinctly before a congressional committee in 1945 that "radiation death is a very pleasant way to die."[20]

In the gradual consolidation of the science—government relationship, begun in 1939 and cemented in 1942, the terrible ethical dilemma of using science as a vehicle for destruction also began to concern the participants. While the nuclear scientists were horrifed by the theoretical implications of the atomic weapon they were creating, most of them would not have given up the opportunity to continue the work.

> We felt that it was a necessity; we were at the beginning of a war where you think more of what is necessary to defeat that enemy of the world, Hitler, and therefore we were not much thinking of the deeper moralities of it.[21]

Victor Weisskopf, the young nuclear physicist who had recently arrived in the United States, later described more personal reasons:

> But to be frank, there were other reasons too. I was a pretty young fellow then; I was not a leader of physics and there were great people in physics working on the project. There was a strong urge to participate with them in this great experience [of] releasing the cosmic forces hidden in the nucleus.[22]

Because of the magnitude of the scientific research tasks, there was an organizational need to coordinate the basic research and development activities. In late 1942, various ad hoc committees appointed by Roosevelt were grouped together as the Manhattan Engineer District, under the direction of General Leslie Groves—a "pushing and bullying"[23] engineer noted for his work on the construction of the Pentagon. The activities of this umbrella organization, were top secret (even the U.S. State Department was unaware of its true purpose until after the bombs had been dropped on Japan).

The "Manhattan Project" was the general name given to the task of developing and building an atomic bomb for use in wartime. Over $2.2 billion dollars were spent on its activities. In December, 1942 Enrico Fermi accomplished the first of the three tasks needed to develop the atom bomb when he created the first controlled chain reaction at the University of Chicago. Now, as Groves has written, "the government was increasingly committed to the ultimate use of the weapon to end the war earlier."[24]

The Manhattan Project

Under General Groves' tough leadership, the necessary machinery for constructing the bomb was geared up for action by the end of 1942. To avoid sabotage and preserve secrecy, Groves scattered the Manhattan Project's installations across the United States and Canada.[25] (By the time Groves finally transferred the project to the newly-created Atomic Energy Commission in January, 1947, there were 37 installations in nineteen states and Canada, with 38,000 contract employees, 3,950 government workers, and almost 2,000 military personnel attached to the project.)[26]

General Groves was responsible for "correlating the entire effort and keeping it directed to the military objectives."[27] He had to procure the raw uranium ore (from the Congo, Colorado, and Canada); insure that the scientists and engineers were adhering to fissionable materials production schedules; and oversee security across the project's network.

When the Allies invaded Italy and Sicily in 1943, General Groves' huge project was, somehow, on schedule. During this time, the small number of large uranium and plutonium industrial production plants (located in Tennessee and Washington) were operated by private industries such as the DuPont Company.[28] Extensive experimental work on chain reactions and isotope separation was being conducted at the University of Chicago, the Argonne Laboratory (near Chicago), Berkeley, Columbia, Princeton, and other universities.

The Bomb Factory: Los Alamos Scientific Laboratory

In early 1943, Groves made two important decisions. First, he chose Los Alamos, New Mexico[29] as the location for the critical "last stage" of the Manhattan project. The Los Alamos facility would grow from a few dozen men in 1943 to over 5,000 people in early 1945. As columnist Drew Pearson stated in his diaries, it was "a unique experiment in complete governmental control."[30]

Second, in March of 1943, he chose the compassionate and brilliant physicist J. Robert Oppenheimer to be the director of the Los Alamos Scientific Laboratory (LASL) located in north central New Mexico on the Parajito Plateau. According to Groves, Oppenheimer, who held joint appointments in the physics departments at Berkeley and Cal Tech, had two disadvantages as director: he was not a Nobel Prize winner and he had no administrative experience.[31]

However, the directorship appointment turned out to be a propitious opportunity for Oppenheimer. "Oppy," as he was called by his colleagues, instilled in the scientific community a camaraderie and a sense of collegiality that compensated for the Spartan living conditions and military atmosphere. Oppy was able to shape his scientific staff (with people from major research universities across the United States, including Berkeley, Cal Tech, Princeton, Chicago, Wisconsin, and Minnesota) into a disciplined research force. Given the critical time limits at Los Alamos there was a need for someone who could get "the utmost in collaboration between civil engineers, metallurgists, chemists, physicists, and military officers."[32]

Oppenheimer, who had lived in New Mexico in the 1920s and had helped General Groves select the Los Alamos site, knew the value of keeping the scientists together in such an isolated setting:

> We needed a central atomic weapons laboratory devoted wholly to work on the bomb, where people could speak freely with each other, where theoretical ideas and experimental findings could effect each other.[33]

Oppenheimer created seven separate work divisions at Los Alamos: theoretical physics, experimental nuclear physics, chemistry and metallurgy, ordnance, bomb physics, explosives, and advanced development division. Until a few months before the actual detonation in July, 1945, much of the work was done without the needed uranium, U-235. While large-scale uranium and plutonium processes were being developed to manufacture these fissionable materials at Clinton, Tennessee, and at Hanford, Washington, very little was available at Los Alamos.[34]

Beyond his good management of the practical tasks associated with building the bomb, Oppenheimer led the scientists to a Hegelian transcendence of their ethical dilemma. Victor Weisskopf recalled that

> The A-Bomb was to be used for destruction, but at the same time Oppenheimer created a spirit, an atmosphere of enthusiasm, of collaboration and of great intellectual and moral purpose. These two things are self-contradictory, but in a way it *was* that way.[35]

Oppenheimer was the beacon of rationality in an isolated, windblown outpost. Life at Los Alamos was "lousy, characterized by isolation, security restrictions, Spartan living conditions, and monotony."[36] Beyond the boredom, the housing shortages, poor roads, scarcity of water, salary complaints, and irregular food supplies,[37] the very real and terrifying "specter of predetonation" was "hovering in the background."[38] Ironically, not until the war's end did the scientists discover that there had been no truly competitive race with the Nazis to be the first to harness the atom for military purposes. Unknown to the Los Alamos scientists, Hitler had cast the fate of his Third Reich with rocketry rather than with the atom bomb. While the German physicists continued their research on atomic fission, they did not have the financial support necessary to develop a super bomb.

Trinity: The First Atomic Explosion, July 16, 1945

Germany surrendered unconditionally to the Allies on May 8, 1945, and the terrible war in Europe was over. Ironically, the war that the Allies fought against Hitler ended before the physicists at Los Alamos and Kaiser Wilhelm Institute in Berlin concluded their attempts to be the first to use atomic power for military purposes. (The Japanese were not seriously engaged in atomic research.) With Nazi Germany's total defeat, some scientists felt that the Herculean efforts to control atomic power for military purposes should come to an end. For some of the European emigre scientists, and others, the moral imperative for working on the atom bomb lost its sharp focus after the death of Hitler.

However, work on the bomb continued and expectations were heightened because nuclear theory was ready to be tested. By July, 1945, the Los Alamos scientists and technicians were ready for the intial atomic test, at the "Trinity" site. The equipment and personnel were carefully moved to Alamogordo, in southern New Mexico, where the historic shot was scheduled.

The "Trinity" bomb was an implosion device which was theoretically more efficient than a gun-type atomic bomb. Technically, however, the detonation of this device was far more complex. In theory, the implosion and gun-type devices would set in motion a process that in milliseconds would culminate in a nonreversible chain reaction, and release vast amounts of energy in the form of heat, blast, and radiation.

The gun-type bomb set off a chain reaction by shooting a particle of uranium down an eight- to ten-foot tube inside the bomb, into a mass of uranium, at a velocity of 2,000 feet per second. ("Little Boy," the bomb dropped on Hiroshima in August, 1945, was a gun-type bomb.)

An implosion bomb, called "Fat Man" by the scientists, was the type detonated at Alamogordo and dropped on Nagasaki. It triggers the chain reaction by suddenly reducing the volume of fissionable material, whether uranium or plutonium. Explosives then compress the material symmetrically and with great force,[39] setting off the reaction.

Great apprehension accompanied the excitement of the scientists at the Alamogordo site since neither detonation process had been tested. "Most of those present were praying—and praying harder than they ever prayed before," said a military officer at the test site.[40] The War Department press release understated the situation when it noted, on July 16, 1945, that "tension before the actual detonation was at a tremendous pitch. Failure was an ever-present possibility. Too great a success, envisioned by some of those present, might have meant an uncontrollable, unusable weapon."[41]

The blast occurred at 5:30 a.m., after hours of delay due to bad weather:

> From the darkness, came the light; a stupendous burst of fierce light many times more brilliant than the noonday sun. Everyone was shocked, dazed, awed, and somehow both pleased and terrified. Something fundamental and primordial struck everyone present.[42]

Kenneth Greisen, who had worked on the implosion bomb, exclaimed, "My God, it worked!"[43] Victor Weisskopf recently recalled the "most grotesque and uncanny feeling I ever had."

The countdown began, 10–9–8, and you could hear over the loudspeaker music from the local FM station. It was playing the Nutcracker Suite by Tchaikovsky. Whenever I hear that melody the whole thing comes back to me. I was over-awed by this terrible sight, by the whole complex of problems that will come with all this and, of course, I was impressed by the great success we had—there was a feeling of pride and distress.[44]

While Oppenheimer's "face relaxed into an expression of tremendous relief,"[45] he felt personal grief mixed with the pride of an awesome scientific accomplishment. After Trinity, he sadly noted, "We scientists have learned sin."[46] A *New York Times* reporter commented that he felt as though he had been "privileged to witness the birth of the world and to be present at the moment when the Lord said, 'Let there be light,'"[47] but Dr. Kenneth Bainbridge, the scientist in charge of the actual Trinity detonation, said simply: "Now we are all sons of bitches."[48]

Less than a month later, the two atom bombs, one gun-type and one implosion, were dropped over Japan, resulting in immense suffering and destruction.[49] The war in the Pacific was over.

Almost immediately after the end of the war, a heated controversy developed among scientists and military and government leaders regarding the future status of atomic power in American strategic military planning. In order to understand the scientists' concerns, it is important to understand what happens when an atomic bomb explodes.

When An Atom Bomb Explodes

Fiery flash. A white-yellow sunflower loudly bursting from its core. The earth-shaking, mushrooming explosion of an atomic bomb is at once extremely violent, frightening, and exhilarating. The immense power of an atomic bomb, totally unlike that of a conventional bomb,[50] is generated by the splitting (fissioning) of uranium or plutonium and the dramatic conversion of this mass into energy—a phenomenon that was, in the past, unimaginable.

Over eighty percent of the energy produced by a bomb blast is in the form of searing, unmerciful heat that reaches a maximum temperature, in a millisecond, of several million degrees centigrade (compared to the 5,000 degrees centigrade generated by a conventional bomb).[51] Within that millisecond, atoms are produced which emit strong gamma-ray radiation—the most dangerous form of penetrating radiation known to man. Finally, trapped within the explosion's fireball is the residual radioactivity consisting of vaporized fission materials. After the explosion, about ten percent of the thoroughly irradiated materials attach themselves to particles in the fireball

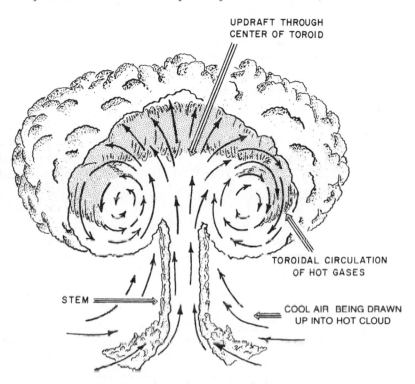

UPDRAFT THROUGH
CENTER OF TOROID

TOROIDAL CIRCULATION
OF HOT GASES

STEM

COOL AIR BEING DRAWN
UP INTO HOT CLOUD

Figure 2. During an atomic explosion, an atomic device, detonated at or near ground level, picks up tons of debris (dirt, pulverized rock, etc.) along with the cool air and the metal bomb casing. The debris, metal, and air are drawn into the stem and brought up into the center of the mushroom cloud. They are then carried aloft and deposited, often hundreds of miles away, in the form of radioactive fallout. (Source: Glasstone and Dolan, *The Effects of Nuclear War*, p. 29.)

itself, are caught in the prevailing winds, and are eventually set free after cooling down, to return to Earth up to thousands of miles away from the explosion, as radioactive fallout.

Of the four effects of a nuclear detonation—blast, heat, immediate nuclear gamma-ray radioactivity, and the dormant radioactivity in the nuclear cloud—it is the residual radioactivity that has generated the most controversy and has posed the many difficult legal, scientific, political, and moral questions that are the heart of the present debate about nuclear testing in America.

While radiation was a known hazard,[52] the danger of radioactive fallout, according to General Groves, "had not received any attention until about January, 1945, when one of the Los Alamos scientists, Joseph Hirschfelder, brought up the possibility that radioactive fallout might be a real problem."[53] At Alamogordo, Groves voiced some concern about the possibility of New

Mexico residents coming in contact with the fallout. But the scientists and the military personnel did not consider fallout to be a major problem. In 1945, Groves and others believed that radiation was "like x-rays." If someone received a heavy dose of radioactivity, said Groves, "he just [took] a vacation away from the material and in due course of time he [would be] perfectly all right again."[54]

When a nuclear bomb is detonated high in the atmosphere, the smaller radioactive particles attach themselves to the bomb casing, which, at the instant of detonation, vaporizes. Upon cooling, these metal fragments, coated with the fission particles, condense to extremely small flakes, remain suspended in air for long periods of time, and come down with the rain and snow thousands of miles from the explosion. A surface detonation, or a low altitude air explosion of an atomic bomb, is another story. When the fireball produced by the detonation contacts the ground, tons of earth, rock, and other matter are lifted into its vortex and become contaminated by combining with the bomb's radioactive fission particles. When the cloud cools, the contaminated debris descends to earth as radioactive "fallout," only a few hundred miles from the detonation. The "fallout" continues to emit radioactive energy in the form of alpha, beta, or gamma rays.

Where the fallout will occur and how much of it will return to the ground depends on several variables: the size of the bomb (how much energy did it produce?), the height of the point of detonation above ground (did the fireball return to the ground?), meteorological conditions (in which direction and how strong was the wind blowing?), the earth under the bomb's fireball at detonation (what was vaporized?), and the incompletion of fissioning that occurred in a particular detonation.

As many as three hundred different types of fission particles can be produced by an explosion, each with a different half-life, that is, the time it takes for the particle's radioactivity to decrease by half. Some of these atoms decay rapidly—within seconds to hours—into a stable (nonradioactive) state. Others, such as iodine–131, have a half-life of 8.05 days, while strontium 90 has a half-life of 28 years. These radioactive particles, or isotopes, "decay" because they emit beta particles (energy) fairly rapidly, and once depleted, become stable.

There are some particles that are not completely split into smaller energy parts. At the moment of detonation, incomplete fissioning of the uranium or plutonium particles in the explosion occurs, resulting in the production of U-233 (with a half-life of 162,000 years) or plutonium 239 (with a half-life of 24,000 years). There is a third result of the detonation itself: neutron-induced radioactive materials are produced—air, rocks, soil, other debris—which have been bombarded by the energized fissioned and unfissioned radioactive elements.

The radioactive particles produced by the complete or incomplete fissioning process are alpha, beta, or gamma types. Alpha particles are very fast moving particles of high energy that carry a positive electric charge. Attached to soil, rock, and other small pulverized debris, they have little penetration power and can go only small fractions of a millimeter into soft tissue. Beta particles carry small negative electric charges and can penetrate only about a centimeter into soft tissue. Gamma-ray coated debris is the most dangerous. These radioactive particles have great penetrating powers and can irradiate the whole body. While alpha and beta rays penetrate the human body very slightly, gamma rays penetrate body tissue much like microwaves. As a result, a person can be irradiated internally from an external exposure to gamma radiation.

Living creatures are exposed to radiation fallout in two ways: *external exposure*, i.e., being showered by radioactive particles from a passing radioactive cloud or coming into contact with fallout already on the ground, and *internal exposure*, i.e., inhaling or ingesting radiated particles (or foods or milk from livestock who ingested radioactive particles).[55]

When radiation passes through a human cell, one of four things happens: no damage, death of the cell, damage to the cell that is then repaired by the body, and damage to the cell "such that it survives and multiplies in its perturbated form over a period of years . . . forming a clone of cells that eventually is diagnosed as a malignancy."[56] The "extent of injury to a cell depends upon the type of radiation, the type of body exposure, the particular sensitivity of the exposed tissue or organ, and the amount of radiation exposure."[57]

Acute radiation exposure and high levels of radiation piercing the living organism cause extensive cell damage, leading to anemia, hemorrhage, infection, diarrhea, and finally, death. Gene mutations may occur in future generations if reproductive cells have been exposed to radiation. In addition to fetal abnormalities, the most common late somatic effects of radiation exposure are various forms of cancer. The brain, lungs, thyroid, kidneys, blood-forming tissues, and female breasts are unusually sensitive to ionizing radiation. Case studies of survivors of Hiroshima and Nagasaki are grim testimony to the horrible biological consequences of an atomic explosion over a populated area.

Post-War Atomic Issues

The fighting ended suddenly in the Pacific. Hiroshima was destroyed on August 6, 1945. The Soviets declared war on Japan on August 8, and Nagasaki was destroyed on August 9. Japan agreed to surrender unconditionally on August 14, 1945. President Truman saw the atom bomb "as the greatest thing in history." Truman had no difficulty issuing the order to bomb the two Japanese cities and exclaimed that "the Japanese in their conduct of the war had been vicious and cruel savages, and I came to the conclusion that if 250,000 young Americans could be saved from slaughter the bomb should be dropped, and it was." But he ordered the suspension of further atomic bombing of Japan because, in his words, "the thought of wiping out another 100,000 people was too horrible."[58] Given the fact that the there was *no bomb to drop* in the forseeable future, Truman's belated agonizing was academic. (When David Lilienthal took over as Chairman of the Atomic Energy Commission in January, 1947, his inspection of the AEC facilities unearthed exactly one atomic bomb and no personnel to build more.)[59]

Almost immediately following the Japanese surrender, two fundamental political-ethical questions were raised about the atomic bomb by the nuclear scientists, the Army generals involved in the project, and politicians: (1) How safe *could* the atom bomb's secrecy remain in the forseeable future? Could America prevent its adversaries (especially Russia) from obtaining information about the atom bomb? and (2) who should *control* the future development of atomic power—civilians or military personnel?

Sharp differences abounded as the participants attempted to answer these questions. Military and political leaders strongly maintained that the atom bomb's secrets were safe and that the Russians could not hope to replicate the scientific developments for at least a generation. Some of the scientists who worked on the bomb, however, rejected the political—military picture and urged an immediate international sharing of the atomic secret, before the Soviets uncovered the clue to harnessing atomic energy. Ironically, the idea of sharing atomic secrets was anathema to them when they were confronting Hitler.

General Groves repeatedly insisted that the Russians were a clumsy, backward, and ignorant people who were incapable of manufacturing an atom bomb before the late 1950s—even if they had the blueprints—because he felt they lacked the technical capacity to do so. In addition, Groves argued before congressional committees in 1945 and 1946 and stated to President Truman that, in his opinion, it would take at least 20 years for the Russians to accomplish what the Americans had in four years. According to Groves, the Russians lacked the scientists, the know-how, and the uranium,[60] to build an atom bomb. "I believe," he stated before a congressional committee in 1945, "that we can keep ahead of any other nation in the world for all time to come, if we have secrecy."[61]

Secretary of War Henry Stimpson, disagreed with Groves and stated to Truman that "we do not have a secret to give away—the secret will give itself away,"[62] Reflecting some of the physicists' views on this matter, Stimpson believed that there were no more secrets to be discovered after the creation of the ultimate weapon. In spite of this disagreement, President Truman accepted Groves' view of the nature of the atomic secrets. When Truman heard of Oppenheimer's complaints about America's post-war plans for maintaining atomic secrecy, the President told Secretary of State Dean Acheson that Oppenheimer was a "crybaby."[63]

Many physicists, aware of the work of their Soviet colleagues, were in fundamental disagreement with Groves' assessment of Russian technical and scientific capabilities and with the military's view that the Americans had at least a twenty-year, atom-bomb monopoly. The scientists predicted that the Soviets would have the atom bomb in three years and pointed out the terrible potential of their new creation. "The atom bomb," said Dr. Phillip Morrison, "is not merely a new weapon; it is a revolution in war." Dr. Alvin Weinberg, a nuclear physicist at the University of Chicago who also worked at the Oak Ridge laboratory, told legislators in 1945, "The bombs at Hiroshima and Nagasaki have made the notion of war in modern civilization an obsolete one." The nuclear scientists called for international control of atomic energy with inspections "by contacts between scientists."[64]

In December, 1945, Oppenheimer reflected the concern of many scientists when he stated before a congressional committee that

> Last summer, the progress of science turned up something very indigestible: The Atom Bomb. . . . We must secure a permanent peace [but] the solution of the problem created by the development of the atom bomb does not lie within the field of science—this is a political problem, a deep problem of human values and a problem of statesmanship.[65]

According to Weisskopf, some of the nuclear physicists were so repulsed by the terrible moral consequences of their creation that they left the program. Groves urged Congress to act quickly to avoid "daily losing key people whose services should be retained."[66] However, no amount of persuasion would have convinced many other scientists to stay. Weisskopf was one of many who departed the project because of the events of August, 1945. "Hiroshima changed my life in many respects," he said. "As a scientist, science was no longer esoteric. . . . The bomb showed all of us that problems of life are not that simple. . . . All things, we learned, are connected, the good and the bad. Science can be good; it can also be a force for the bad."[67]

In June, 1945, a group of distinguished nuclear physicists at the University of Chicago, headed by Nobel Prize winner James Franck, prepared a report on a number of issues involving atomic energy for the Secretary of War, Henry L. Stimpson. The Franck Report stated, in part;

> In Russia (as in England, France, and Germany) the basic facts and implications of nuclear power were well understood in 1940, and the experience of the Russian scientists in nuclear research is entirely sufficient to enable them to retrace our steps within a few years, even if we should make every attempt to conceal them. . . . Maintaining secrecy . . . can protect us for [no] more than a few years.

We cannot hope to avoid a nuclear armament race either by keeping secret from the competing nations the basic facts of nuclear power or by cornering the raw materials required for such a race.[68]

In their conclusion they argued against continuing military use of the atom bomb, so that an international accord on the control of atomic energy could be reached. Simultaneously, hundreds of scientists and technicians, at the universities and at the Oak Ridge and Los Alamos labs spontaneously organized in an effort to explain the danger of atomic power and the fact that the secret was not secret.[69] Einstein stated that he had the "inescapable responsibility to carry to our fellow citizens an understanding of the simple facts of atomic energy and its implications for society,"[70] and publicly argued that

through the release of atomic energy, our generation has brought into the world the most revolutionary force since prehistoric man's discovery of fire. This basic power cannot be fitted into the outmoded concept of narrow nationalisms. For there is no secret and there is no defense; there is no possibility of control except through the aroused understanding and insistence of the peoples of the world.[71]

The scientists, evidently unmindful of value judgments they drew about the comparative evils of Hitler and Stalin, did not sway the politicians. In 1945, these general beliefs of the scientific community regarding nuclear power, secrecy, and the importance of international control were not accepted by America's military or political leaders. Moreover, the American public was not interested in international control of atomic power.

Once again, the Americans had reluctantly fought, and helped win, another European War. This time, however, things were different. The American atom bomb monopoly might mean that our young men might not have to fight a third war in Europe. Also, by 1946 and 1947, Americans had begun to buy their way out of the Depression.[72] With the fighting ended, the public was now concerned about inflation, buying a new car, the middle-class life, and enjoying security and peace. The public believed the generals and the President when they said that the Soviet Union could not possibly develop the atom bomb. They were not thinking of the newly-emerging international problems created by the awesome new weapon. Americans felt safe with the knowledge that America was the sole possessor of this terrible new weapon.

When Congress created the new civilian agency, the U.S. Atomic Energy Commission, in 1946 to supervise the development of atomic energy, only one of the five commissioners had any kind of scientific knowledge of nuclear physics and atomic energy: Dr. Robert F. Bacher, a Cornell physicist, had worked on the bomb at Los Alamos and had assembled the core of the "Fat Man" bomb at the Trinity site.

The first AEC Chairman was David E. Lilienthal, former head of the Tennessee Valley Authority. He was already on record as strongly supporting Groves' view that the Russians could not possibly build an atomic bomb.[73] The other AEC members were Sumner T. Pike, a former member of the Securities and Exchange Commission; William W. Waymack, a newspaper editor from Des Moines, Iowa; and Louis L. Strauss, a New York investment banker. (General Groves would continue to play a major role in the AEC until his retirement in 1948. He strongly opposed civilian control of atomic energy and convinced Congress to create a Military Liaison Committee to work with the AEC.)[74]

Although Lilienthal was greatly surprised to find that our stockpile in 1947 consisted of one atom bomb, the AEC evinced no great panic or sense of urgency. After all, the Russians did not have the ultimate weapon and the Americans did: it was the age of Pax Atomica. But it was to be a short-lived era of contentment in the United States. During the construction of the atom bomb, conflicts surfaced over the role and function of scientists in a free society—especially when working for the government to construct an instrument of unadulterated terror.

American science, and its scientific community, came down from the ivy-covered university towers to labor endlessly (but with all expenses paid) at developing the atom bomb. Whether it was out of patriotism or the desire to work with brilliant Nobel Prize-winning physicists on the scientific challenge of the century, most nuclear physicists set aside the moral contradictions and worked to develop the bomb.

Generally, scientists are deeply, almost mystically, committed to the notion of sharing ideas. (Recall the German physicist Hahn, in late 1938, who shared his discovery with his exiled, Jewish-Austrian physicist colleague and friend of 30 years, Lise Meitner.) As academics, except for the European emigré scientists working on the atom bomb, the nuclear scientists never felt entirely comfortable with the secrecy that pervaded Los Alamos. However, during the War none of them was prepared to share these atomic secrets with their Nazi counterparts who, it was believed, were working feverishly at the Kaiser Wilhelm Institute in Berlin to uncover the same secrets. Indeed, for all the physicists at Los Alamos, there was the shared belief that they were in a "life-and-death-of-civilization" race with their former colleagues toiling in Hitler's Germany.

After the war, the feeling of the nuclear science community regarding atomic openness and secrecy was different. Perhaps because of the ultimate nature of their discovery, combined with their high hopes for world peace and their belief that the Soviets would soon create their own atomic bomb, many scientists felt that it was imperative that scientific openness be fully restored. This idealism, however, was not the policy of the American government after World War II. New conflicts surfaced again when the Atomic Energy Commission, the nation's "atomic shield,"[75] began its work in 1947.

2 The Atomic Energy Commission and American National Security, 1946–1958

Fallout is certainly all right, if you don't live next door to it.
Lewis Strauss, Chairman, AEC

How dare they decide to protect their children living in populated areas at the expense of our children.
Gloria Gregerson, Bunkerville, Nevada

We must not let anything interfere with [these] tests—nothing.
Thomas E. Murray, AEC Commissioner

The Atomic Energy Commission: Its Creation and Mission

The Trinity shot at Alamogordo, New Mexico, in July, 1945, ushered in the Nuclear Age. Congress was not in session when the atom bombs were dropped over Hiroshima and Nagasaki on August 6 and 9, 1945. After reassembling on September 5, 1945, the 79th Congress was confronted with a critical question: who would control the future development and deployment of this awesome new weapon of war—the military or the civilian authorities?[1]

In the space of four hectic months, September through December, 1945, fifty bills were introduced in the House and Senate that focused on the question.[2] For President Harry S Truman, and many legislators, it was important that the primary responsibility for the development of atomic energy and the control of the atomic arsenal rest with the civilian authorities. The President said: "I regard the continued control of all aspects of the atomic energy program, including research, development and the custody of atomic weapons as the proper functions of the civil authorities."[3]

The director of the Manhattan Project, General Leslie Groves, would not have gone so far. He and others in the military establishment, along with some in the Truman administration and Congress, strongly believed that the military should exercise major control over the development of atomic

20

power and the use of the bomb.[4] Groves and other military men supported the Congressman Andrew May–Senator Edwin Johnson bill, which, if passed, would have established a part-time civilian staff with full-time military officers serving as adminstrator and deputy administrator of the atomic power agency. The bill reflected May's view: "The War Department discovered the weapon. Why can they not keep the secret?"[5]

The debate over May—Johnson raged in Congress for many months. Hearings were held in Washington, D.C. to explore the basic questions relating to the development and control of atomic energy.[6] In late fall, 1945, members of Congress, scientists (who were concerned about the heavy security measures in the May–Johnson legislation and the stiff penalties for violation),[7] and members of President Truman's staff blocked the Army's attempt to push the legislation through.

On December 20, 1945, freshman Senator Brian McMahon (D-Connecticut) introduced Senate Bill 1717, which was used by Congress in approving the Atomic Energy Act, and ultimately signed into law by President Truman on August 1, 1946.[8] Given the compromise character of legislative policymaking, the McMahon legislation provided for the military to oversee atomic development, but for civilians to have ultimate control. While civilians would have the main responsibility for developing, using, and controlling atomic energy in the United States, provisions were built into the bill for military review of AEC field activities.

The first section of the Atomic Energy Act spelled out the policy of the people of the United States, which was to develop, use, and control atomic energy "for the public welfare, *subject at all times to the paramount objective of assuring the common defense and security.*" The 1946 Act, Section 2(a), created an Atomic Energy Commission (AEC), which consisted of (1) five civilians nominated by the President and confirmed by the U.S. Senate; (2) a General Advisory Committee, consisting of nine persons selected by the President for six-year terms, to advise the AEC on scientific and technical matters relating to research and development; (3) a Joint Committee on Atomic Energy responsible for overseeing the AEC, consisting of nine senators and nine representatives selected by the President Pro Tempore and the Speaker of the House of Representatives; (4) a General Manager to "discharge the administrative and executive functions" of the AEC; (5) a Division of Military Application, to advise the AEC on military applications of atomic energy; and (6) most important, a Military Liaison Committee, consisting of military representatives from the Army, Navy, and Air Force.[9]

The legislation gave these military representatives the important task of "advising the AEC on all atomic energy matters which the Military Liaison Committee deems to relate to military applications, including the development, manufacture, use and storage of atomic bombs." The Atomic Energy Act also gave the Committee the power to go to the Secretary of War (later the Secretary of Defense), if they believed that the AEC failed to act "in a manner appropriate for national security."

Beyond the military needs of manufacturing and stockpiling atomic weapons and the continuous testing of new atomic weapons, the general tasks of the AEC were outlined in the 1946 legislation. The AEC was to assist and foster private research and development; totally control—by state monopoly—scientific and technical information about atomic energy; conduct federally run research and development, and completely control the production, ownership, and use of fissionable material in absolute secrecy "to assure the common defense and security."[10] All these activities were to "be carried on only to the extent that the express consent and direction of the President of the United States had been obtained, at least once a year."

In October, 1946, President Truman appointed five men to serve on the AEC on an interim basis. On January 15, 1947, the five names were sent to the Senate for confirmation and, after lengthy hearings, the first set of AEC commissioners was confirmed by the U.S. Senate on April 9, 1947.[11] Led by their chairman, David E. Lilienthal, and supported by their newly appointed and confirmed civilian General Manager, Carroll Wilson, the AEC commissioners—Lilienthal, Robert F. Bacher, Lewis L. Strauss, Sumner T. Pike, and William A. Waymack—began their formal full-time commission work in the spring of 1947.

Exactly ten days after his Senate confirmation, in a speech before the American Society of Newspaper Editors, on April 19, 1947, Chairman Lilienthal said that "decisions on atomic energy should not be made in secret. . . . I am proposing a broad and sustained program of education at the grass roots of every community in the land."[12] For David E. Lilienthal, the new AEC chairman, and for other AEC commissioners, at least one aspect of their new mandate—the idea that the nation's atomic energy program would be run in absolute secrecy—was clearly troublesome. But if their initial move was to challenge this aspect of America's postwar nuclear program, within a short period of time their position had changed. Influenced by the political, military, and technological events in the world, and by the contradictions surrounding the atomic energy issue, the views of these five commissioners "inexorably shifted" from a belief in openness and the peaceful use of the atom to the "grim realization that for reasons of national security atomic energy would have to continue to bear the image of war."[13]

When the AEC commissioners formally assumed their responsibilities, as outlined in the Atomic Energy Act of 1946, they were confronted with three major technical problems. The first was to build up the depleted nuclear stockpile by re-establishing and maintaining the production (mining) of raw materials and the manufacture of fissionable materials.

Second, the AEC had to deal with the scientific "brain-drain," from the military's atomic research and development operation, which had been going on since the war's end. General Groves faced the dramatic emotional resignation of Oppenheimer (who had told Truman, "Mr. President, I have blood on my hands") and the resignation of the scientific staff at Los Alamos."[14] Groves made vital improvements to the Los Alamos facilities in an attempt to keep the scientists who were leaving for practical rather than moral reasons, and Oppenheimer's replacement, Norris E. Bradbury, attempted to "establish conditions as nearly ideal as possible" to halt the further loss of scientific talent.[15]

However, the staff "melted away," and, in 1946, according to Groves, the lab was "in a crisis."[16] In testimony before the Joint Committee on Atomic Energy in February, 1949, Lilienthal listed the "recruitment of scientific and technical talent" as an "important research and development problem yet to be solved."[17]

Finally, the AEC commissioners faced the administrative nightmare of reorganizing and managing all of the facilities and forces under its jurisdiction. In 1948, the AEC "owned more real estate, plants, and equipment than General Motors; ran more buses than the city of Philadelphia; and had land holdings larger than Rhode Island."[18] The Agency had laboratory facilities in Washington and Tennessee; administrative offices in New York and Washington, 38,000 private contractors and employees involved in manufacturing and research; and scientists in dozens of universities spread across the country doing theoretical work in nuclear physics.

An adminstrative organization that would implement the necessary changes efficiently was imperative. What the commissioners decided was to set up a headquarters of operations "that was informal, flexible, and free from the incubus of a cumbersome staff."[19]

Lilienthal and the other commissioners, in 1948, made a fateful decision about managing the far-flung AEC operations. The AEC commissioners decided to act exclusively as a policy board with the General Manager providing the AEC with judgment and information from five other managers across the country.[20] The touchstone of this management operation was decentralization. The five field managers were Carroll Tyler, in charge of atomic research and development and the production of atom bombs, Los Alamos, New Mexico; John Franklin, in charge of the production of fissionable material and research and development on atomic energy, New York; Wilbur Kelley, in charge of the production of uranium–235 at Brookhaven National Laboratory, New York; Alfonzo Tammaro, in charge of development at the Argonne National Laboratory, Chicago, and at Hanford, Washington; and Carleton Shugg, in charge of the manufacturing of plutonium and the management of the Pacific tests.[21] Each field manager was given "a large share of the detailed administrative controls"[22] under the new AEC operating set-up.

These major responsibilities were delegated to the field managers at these far-away AEC facilities, and the five men were required to report only to Washington, D.C. on matters involving AEC policy. Completely free to hire and fire personnel at their facilities, the managers could negotiate contracts with private contractors and, critically important, they could "*issue their own directives on how AEC directives should be met*"[23] [italics mine]. The AEC's historian noted that the Washington, D.C. office had no line responsibility "under this decentralized process."[24]

Critics, including members of Congress, some scientists, and all nine members of the AEC's General Advisory Committee, viewed Lilienthal's effort to decentralize the AEC as nothing but "incredible mismanagement."[25] Lilienthal's decentralizing tendencies led to a major investigation by Senator Bourke B. Hickenlooper (R-Iowa) in May of 1949 that, according to the Senator, focused on "the administrative direction and policies of Mr. Lilienthal as Chairman of the Commission itself."[26] There were insistent calls for Lilienthal's resignation. Soon, however, global events dwarfed the AEC's controversy. Soviet challenges to the free world in the late 1940s and early 1950s led to two major American decisions: to construct a "super bomb" and to develop a major atomic weapons testing facility in the continental United States: The Nevada Test Site.

By 1947, relations between the United States and the U.S.S.R. had changed considerably. In response to Soviet takeovers in Eastern and Central Europe, America developed the Truman Doctrine and the Marshall Plan, which were largely successful in providing Europe with recovery support. By early 1948, over 70% of Americans viewed Russia unfavorably. In the spring of 1948, in violation of agreements made at Potsdam in 1945, the Soviets blocked America's entry into Berlin. For over a year, the United States and its allies defied the Berlin blockade and flew 1,017 daily flights into the city, around the clock, to supply its 2.5 million residents with the necessary food and fuel. In May, 1949, the Russians relented and allowed vehicular traffic to move into Berlin's occupied zones.[27]

In September, 1949, President Truman announced that an "atomic explosion" had occurred in the Soviet Union. His choice of words may have been an attempt to distinguish between a device and an operational bomb, or may have reflected Truman's view (shared by Groves and other military leaders) that the Soviets were scientifically inferior and that the blast was probably the result of an accident rather than a deliberate detonation.[28] The Russians, however, had exploded an atomic device. Military and political leaders responded to this realization (scientists were not surprised) by expanding the AEC's plant construction and accelerating the manufacture of atomic bombs. In addition, there was a renewed—but this time successful—call for the development of a new, more powerful nuclear weapon. A troubled Congress funded the expansionist efforts of the AEC. (The AEC's 1950 appropriation was $1.5 billion; in 1951, due to the AEC's commitment to build more atomic bombs, to begin work on the super bomb, and to expand atomic research and development facilities, the AEC's budget leaped to $2.75 billion.)[29]

The concept of a super bomb—a fusion-hydrogen bomb—had existed since the early days of the Manhattan Project. In 1943, Edward Teller, one of many Jewish emigré physicists from Nazi-occupied Europe working at Los Alamos, had begun the theoretical development of the new bomb. After the war, both Teller and Ernest Lawrence, the Nobel prize-winning physicist from Berkeley, continued to push for further exploration of this new nuclear weapon[30] that was a hundred times more powerful than the atom bomb. Teller and Lawrence believed that the super bomb could be built and should be built in order for the United States to maintain nuclear superiority over the Soviets. However, until the Soviets exploded their atomic bomb in 1949, there was very little support for the development of this nuclear super-weapon, soon to be called the "hydrogen bomb."

Even after the explosion of Russia's atomic bomb in 1949, some of the AEC commissioners and all of the AEC's General Advisory Committee, chaired by Robert Oppenheimer, argued against developing this new weapon. A crash program to develop the hydrogen bomb posed new moral and technical questions for these scientists. Said Oppenheimer: the bomb "is a weapon of genocide."[31] (But in 1951, in another moment and in another context, Oppenheimer would refer to the hyrogen bomb as "sweet and lovely and beautiful.")[32]

The new chairman of the AEC, Gordon Dean (Lilienthal submitted his resignation to Truman in 1949), was also opposed to the development of the new bomb. He urged Truman to enter secret diplomatic negotiations with the Russians to seek international control of atomic energy rather than have America unilaterally escalate the nuclear arms race.[33] However, Lewis Strauss suggested in a letter to his fellow AEC commissioners, that it "was time for America to build a 'super' bomb."[34] He proposed that the AEC establish a crash program for the development of the new "super bomb." Despite the scientific community's general opposition to the development of the nuclear weapon, the die was cast for its development. By the end of 1949, after further revelations about a major spy scandal that shocked the nation, America was committed to the development of the hydrogen bomb. Within a few years, the hydrogen bomb would be successfully tested in the Pacific, followed shortly by the Russians' successful test of their hydrogen bomb.[35]

Americans were certainly frightened by the revelation that the Russians had the secret of the atom bomb. In 1949, Americans also discovered that the Russians had received information about the Manhattan Project from spies who had worked at Los Alamos and other atomic energy research and development laboratories during the war.

The "betrayers of our atomic secrets"[36] were revealed to a shocked Amer-

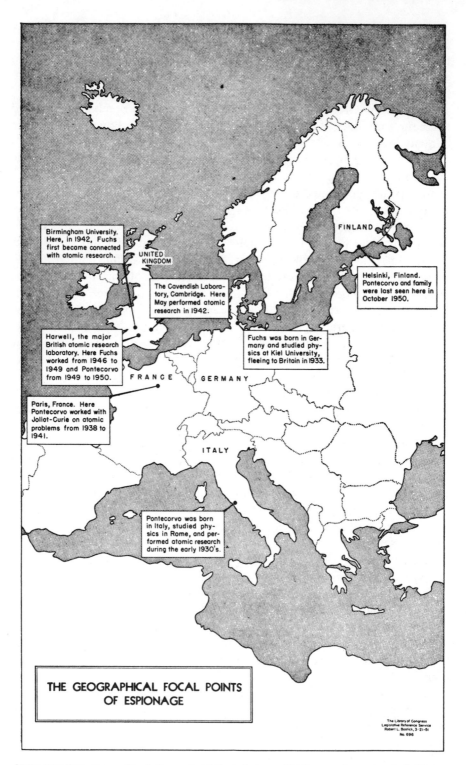

Birmingham University. Here, in 1942, Fuchs first became connected with atomic research.

UNITED KINGDOM

FINLAND

The Cavendish Laboratory, Cambridge. Here May performed atomic research in 1942.

Helsinki, Finland. Pontecorvo and family were last seen here in October 1950.

Harwell, the major British atomic research laboratory. Here Fuchs worked from 1946 to 1949 and Pontecorvo from 1949 to 1950.

Fuchs was born in Germany and studied physics at Kiel University, fleeing to Britain in 1933.

FRANCE   GERMANY

Paris, France. Here Pontecorvo worked with Joliot-Curie on atomic problems from 1938 to 1941.

ITALY

Pontecorvo was born in Italy, studied physics in Rome, and performed atomic research during the early 1930's.

THE GEOGRAPHICAL FOCAL POINTS OF ESPIONAGE

The Library of Congress
Legislative Reference Service
Robert L. Bostick, 3-21-51
No. 696

Source: Joint Committee on Atomic Energy, *Soviet Atomic Espionage,* 82d Congress, 1st session, April, 1951, pp. vi–vii.

ican public. The list included Dr. Klaus Fuchs, a German-born British scientist. Klaus, who worked at Oak Ridge (1943), Columbia University (1944), and Los Alamos (1944–1946), and in 1946 was the Chief of the Theoretical Physics Division at Harwell (Britain's Los Alamos equivalent), had been passing atomic secrets to the Soviets since 1943. For the congressional committee investigating this spy scandal, it "was crystal clear that Fuchs was the most damaging of the spies" for he was privy both to atomic and hydrogen bomb experiments and to theoretical discussions.[37]

Maps of "The Geographical Focal Points of Espionage" in Europe and the United States were prepared by congressional staffers to graphically depict the worldwide communist effort to gather information about the U.S. atomic bomb development during and shortly after World War II.

Spy charges were brought against Dr. Allan May, a Canadian-born British physicist; Dr. Bruno Pontecorvo, an Italian-born British scientist; and an American, David Greenglass, an Army NCO who had been stationed at Los Alamos in the summer of 1944. Greenglass was accused of passing atomic information to Harry Gold, an "espionage courier," as he was labeled by the Congressional Report, who relayed information from Fuchs and Greenglass to Soviet officials in America.

Also charged with "espionage activities" were Julius and Ethel Rosenberg (who was David Greenglass's sister). The Rosenbergs were convicted in March, 1951, and were executed by the federal government the following year.

According to the legislative investigation, the information given to the Russians "advanced the Soviet atomic energy program by eighteen months at a minimum. . . . A major share of [the Manhattan Project experiment] was at hand for Russia to exploit without the independent exertion on her own part otherwise necessary."[38]

By June, 1950, with the outbreak of hostilities between North Korea and South Korea, America and Russia became locked in the nuclear arms race that continues today. A recent U.S. Defense Department document stated "The origin of all United States nuclear test series can be traced to the post-World War II tension between the United States and the Soviet Union. . . . Expecting eventual Soviet development of nuclear weapons, the United States continued to expand its nuclear arsenal to maintain superiority over its most potentially dangerous adversary."[39]

Although the AEC had considered conducting atomic testing on the American continent before the Korean hostilities began, America's nuclear testing range was the Pacific Ocean. The advantages of testing atomic devices on land close to the scientific and technological centers at Los Alamos and other locations were obvious. Such a domestic testing site would reduce the "weapons development lead time and considerably less expense would be incurred."[40] However, in March, 1949, AEC Commissioner Sumner Pike noted that "only a national emergency could justify testing in the United

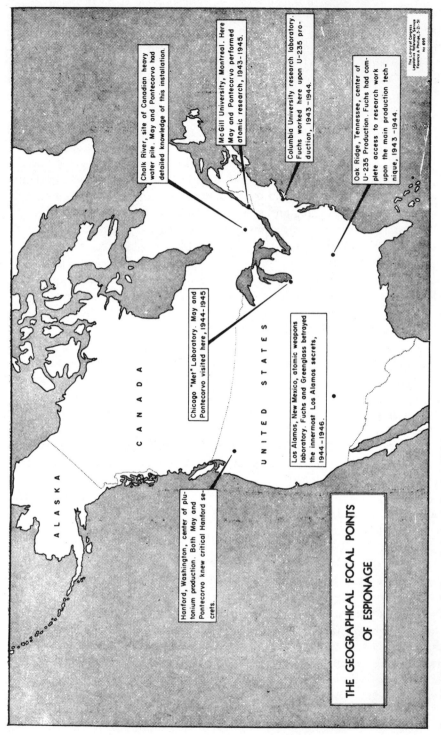

Chalk River, site of Canadian heavy water pile. May and Pontecorvo had detailed knowledge of this installation.

McGill University, Montreal. Here May and Pontecorvo performed atomic research, 1943-1945.

Columbia University research laboratory. Fuchs worked here upon U-235 production, 1943-1944.

Oak Ridge, Tennessee, center of U-235 Production. Fuchs had complete access to research work upon the main production technique, 1943-1944.

Chicago "Met" Laboratory. May and Pontecorvo visited here, 1944-1945

Los Alamos, New Mexico, atomic weapons laboratory. Fuchs and Greenglass betrayed the innermost Los Alamos secrets, 1944-1946.

Hanford, Washington, center of plutonium production. Both May and Pontecorvo knew critical Hanford secrets.

ALASKA

CANADA

UNITED STATES

THE GEOGRAPHICAL FOCAL POINTS
OF ESPIONAGE

The Library of Congress
Legislative Reference Service
Florence A. Phillips, 3-21-51
No 695

Source: Joint Committee on Atomic Energy, *Soviet Atomic Espionage,* 82d Congress, 1st session, April, 1951, pp. vi-vii.

States."[41] Less than a year earlier the AEC commissioners had rejected a continental United States testing site because the "physical problems and domestic political concerns were too complicated to warrant [its] creation."[42] But when the Korean War began, the AEC immediately moved for a continental American testing site. In the AEC's perspective, Korea became the emergency that justifed a continental testing site.

The AEC supported the concept of a continental test site in order to maintain its nuclear weapons superiority. America needed to continue atomic testing at a location "where its basic security and general accessibility [would not] be jeopardized by enemy action."[43] On December 18, 1950, President Truman granted permission to the AEC to develop the Las Vegas-Tonopah Bombing and Gunnery Range (U.S. government-owned land north of Las Vegas, Nevada) as America's new domestic atomic test range. (The AEC continued to use the Eniwetok Proving Grounds in the Pacific Ocean for atomic testing during the 1950s.) According to the minutes of an AEC commissioners' meeting, on December 12, 1950, "two fundamental uses" were seen for a domestic test site: (1) as "supplemental to Eniwetok"; and (2) as an "emergency alternate to overseas sites."[44]

At that meeting, the AEC commissioners concluded that no continental site in America could be considered "a completely safe alternative to overseas sites."[45] Due to "safety considerations, i.e., radiological hazards," the domestic site would primarily be used for atomic tests "involving relatively low orders (up to 50 kilotons) of energy release."[46] The Nevada site was finally selected over Dugway, Utah and Alamogordo, New Mexico because of its radiological safety features: low population density, favorable meteorological conditions (a prevailing easterly wind blowing away from the populous west coast), and good geographical features, i.e., hundreds of miles of flat, government-controlled land.[47]

The Nevada Test Site became the primary location for American testing of nuclear weapons - especially tactical weapons systems developed by AEC and Department of Defense technicians, scientists, and engineers. According to Mahlon E. Gates, Manager of the Nevada Operations Office of the U.S. Department of Energy, between 1951 and 1958, *every* atmospheric atomic bomb test at the Nevada Test Site was "weapons related."[48] On July 16, 1951, the Military Liaison Committee wrote to the Chairman of the AEC, Gordon Dean, requesting that the AEC grant permission to the Department of Defense to use military troops at the Nevada Test Site. The Committee wrote that "indoctrination in essential physical protective measures under simulated combat conditions and observations of the psychological effects of an atomic explosion are reasons for this desired participation."[49] On August 3, 1951, the AEC sanctioned the placement of troops at nuclear tests in the Nevada desert.[50] For the next seven years, tests at the site were often

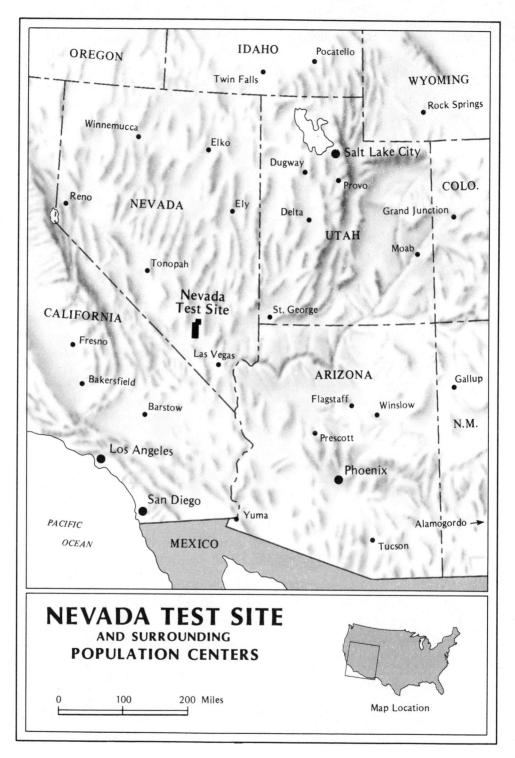

NEVADA TEST SITE
AND SURROUNDING
POPULATION CENTERS

Map Location

```
0        100        200  Miles
```

conducted so that the Department of Defense could "train military units to become familiar with new weapons and their characteristics."[51]

Initially, however, the use of military troops at the Nevada Test Site was a difficult issue for the AEC. The decision to allow them at the site had been made under pressure exerted on the civilian commissioner by the Department of Defense's Military Liaison Committee, despite the troops' potential vulnerability to radiation health hazards in the test field. Then, in the early stages of the testing program, the AEC commissioners acted against the wishes of their medical advisor in the Division of Biology and Medicine, and gave the military permission to establish their own health standards for troops used on the Nevada flats.

At first, the AEC attempted to limit the troops to a distance of seven miles from the detonation site. The military, however, urgently requested that the troops be allowed to move in as close as 7,000 yards (3.9 miles) from ground zero, and, after detonation, to move onto ground zero. Brigadier General A.R. Luedecke, USAF, wrote, "[we] are prepared. . . and desire to accept full responsibility for the safety of all participating troops and troop observers."[52] The AEC's medical director, Dr. Shields Warren, recommended that the Commission deny the request because of the adverse impact on public opinion should some soldiers be injured or killed by the atomic blasts and because each "explosion is experimental and its yield cannot be predicted with accuracy."[53]

Yet, in April, 1952, the AEC approved the military's request. AEC Chairman Gordon Dean wrote a letter, dated April 2, 1952, to the Chief of the Armed Forces Special Weapons Project. In his letter, Dean disclaimed AEC responsibility for any problems that might develop, and requested that "the Exercise Director [military officer] prepare a safety plan to minimize risk of injury which is *acceptable to the Test Manager*. The responsibility for troop compliance with this safety plan rests, of course, with the Exercise Director"[54] [italics mine].

The AEC commissioners knew at the beginning of the test program, in 1951, that the site would be used to develop atomic power for military use and that military troops would occasionally be used in these atomic test exercises. In 1951, Gordon Dean had commented that the testing was done to "develop a situation where we will have atomic weapons in almost as complete a variety as conventional ones. This would include artillery shells, guided missiles, torpedoes, rockets, and bombs for ground support aircraft. . . . We could use an atomic bomb today in a tactical way against enemy troops in the field. . . . We are steadily increasing, through our technology and production progress, the number of situations in which atomic weapons can be effectively employed in battle areas."[55] Yet, the problem of balancing military needs against safety was never resolved by the AEC commissioners.

The Above-Ground Atomic Testing Program, 1951–1958

On January 27, 1951, little more than a month after Truman approved the Nevada Test Site, the initial atom bomb tests began on the southwestern desert. Seven major test series were conducted at the Nevada Test Site from January, 1951 to October, 1958 (when a moratorium on atmospheric testing was agreed upon by the United States and the Soviet Union): (1) *Ranger*, January–February, 1951, (2) *Buster-Jangle*, October–November, 1951, (3) *Tumbler-Snapper*, April–June, 1952, (4) *Upshot-Knothole*, March–June, 1953, (5) *Teapot*, February–May, 1955, (6) *Plumbob*, April–October, 1957, and (7) *Hardtack, II*, September–October, 1958. All of these developmental tests were necessary because weapons systems do not remain static. "The technology utilized constantly changes. . . . All military research and development involves experimentation as a means of verifying theories and concepts."[56] This principle, referred to as the weapons design loop (Figure 3), meant that from its inception, the AEC was pressed to develop a military research and development operation that would quickly provide the Department of Defense and the AEC with feedback about theoretical miscalcula-

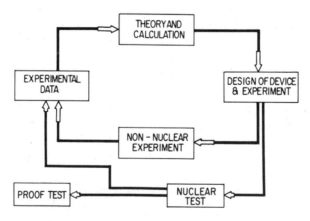

Figure 3. Diagram of a Weapons Design Loop, illustrating the critical interaction between the Los Alamos scientists, Department of Defense technicians, and the Nevada Test Site test director. The former, because they constantly needed to test their atomic devices in order to modify and improve them, were continuously demanding that test shots be detonated on time and that additional shots be added to acquire even more data on the behavior of these atomic devices. The test director was caught between the safety imperative and the demands of the military technicians and the Atomic Energy Commission scientists. (Source: Final Environmental Impact Statement, *Nevada Test Site*, U.S. Energy Research and Development Administration (ERDA), September, 1977, pp. 2–5.)

tions and/or possible design flaws in the atomic weapon they were developing.

The AEC's field office in New Mexico (Los Alamos–Santa Fe) was the "focal point for all NTS testing activities"[57] (Figure 4). Using the weapons design loop concept, the following procedure was followed for each test at the Nevada Test Site between 1951 and 1958.

1. Los Alamos scientists would design a series of atomic tests.
2. The test designs would be submitted to the field manager at the AEC field office in Santa Fe, New Mexico.
3. The proposals would then be sent to the AEC's Division of Military Application in Washington, D.C. for review and approval.
4. If the Division of Military Application approved the proposals, then the AEC commissioners had to approve the test series.
5. After AEC approval, the proposals were sent to the Departments of Defense and State.
6. Approval was obtained from the National Safety Council.
7. Approval was obtained from the President of the United States.
8. The actual detonation authority was given to the field manager, Santa Fe, delegated by the general manager, AEC, Washington, D.C.[58]

Operationally, the chain would begin and end at the local site. The scientists at Los Alamos, and the field manager in charge of the operation, would initiate the series of tests and would complete the test series. The Division of Military Application in Washington, D.C. also played a critical role in responding to the plans of the Los Alamos technicians. At this juncture both the AEC and the Department of Defense, especially the unit responsible for planning for the use and deployment of strategic and tactical atomic weapons, the Armed Forces Special Weapons Project, "shared responsibility for planning and implementing the atmospheric nuclear weapons test program. The AEC was responsible for exploring and developing new areas of nuclear weapons technology, while the DOD was to incorporate the weapons into the military defense program."[59]

After the atomic test series was approved up the line of command, the Nevada Test Site *test manager* became the central planning figure. He was assisted by personnel from the New Mexico office, AEC contractors, and various Defense agencies, including Los Alamos scientists on the Advisory Panel who advised the manager about weather conditions, etc.[60] (Figure 5 illustrates the organization of the Nevada Test Site.) Dr. Gordon M. Dunning, AEC Director of Biology and Medicine, testified that "before each shot a team [the Advisory Panel] of biologists, medical doctors, meteorologists, fallout prediction specialists and blast experts would review weather conditions and recommend cancelling the shot."[61]

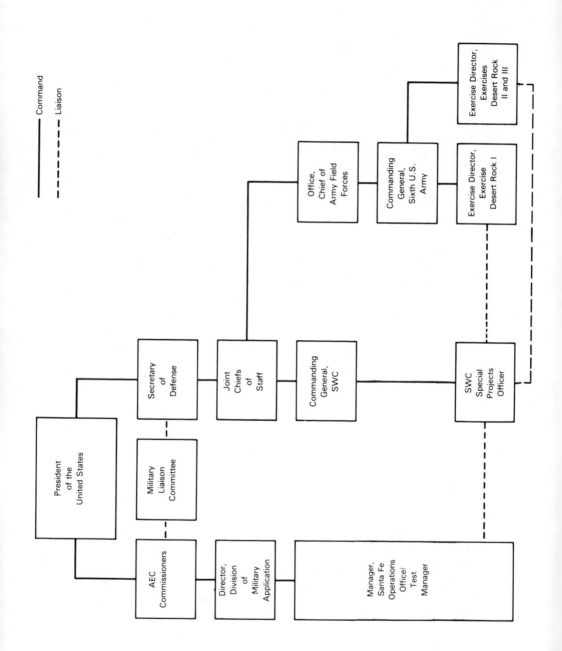

Command

Liaison

President of the United States

Military Liaison Committee

Secretary of Defense

AEC Commissioners

Director, Division of Military Application

Joint Chiefs of Staff

Commanding General, SWC

Office, Chief of Army Field Forces

Manager, Santa Fe Operations Office/Test Manager

SWC Special Projects Officer

Commanding General, Sixth U.S. Army

Exercise Director, Exercises Desert Rock II and III

Exercise Director, Exercise Desert Rock I

The test manager was always under a great deal of pressure to conclude the test shots by a specified time, so that scientists, technicians, and military strategists could advance to the next phase of nuclear weapons development. Because the above-ground nuclear tests were conducted during a period of hot and cold war, America was under great pressure to maintain its superiority over the Russians in the development of nuclear weaponry. Any delay caused by dangerous weather conditions was difficult for the test manager.

On May 21, 1953, for example, at an AEC commissioners' meeting in Washington, D.C., M.W. Boyer, an AEC staff member, asked if the fact that the test organization was under pressure to meet the test schedule made them "more apt to take chances when they are running behind."[62] Mr. Alvin Graves, the test director, responded to Boyer's question by stating that "safety standards are adhered to regardless of whether or not the series is on schedule. Of course, mistakes in judgment are always possible, especially at the end of a very tiring test series."[63] Gordon Dean, the AEC Chairman, asked that "the concern of the Commission" be conveyed to the test organization and that "everything be done to avoid another fallout over St. George" a small Utah town, due east of the Nevada Test Site.[64]

The dialogue between AEC Commissioner Gordon Dean and AEC bureaucrat-in-the-field, Alvin Graves, reflected an important bureaucratic problem: the fundamental differences between a political appointee and the civil servant in the same agency.[65] Test director Graves was compelled to detonate the shots within a specified time period. AEC Chairman Dean, the political appointee, was concerned about political and public responses to testing accidents and about the continuation of high levels of congressional funding for atomic energy research and development, but he was not subjected to the type of "testing" pressure and tension that Graves confronted. The difference was obvious and dramatic. For Dean and the AEC commissioners, politics called for prudence in testing. For the AEC operatives in the field, military urgency and scientific and technological necessity made adhering to test series timetables a major goal.

The minutes of an AEC commissioners' meeting in Washington reported that AEC Chairman Dean

---

Figure 4. The Atom Bomb's Test Organization structure. There would be parallel movement, and authorization given, up the chain of command until the request to test reached the President's desk. He was the final approver of all Atomic Energy Commmission test firings at the Nevada Test Site and elsewhere. (Source: Department of Defense, *Operation Buster-Jangle, U.S. Atmospheric Nuclear Weapons Test Personnel Review*, DNA-6023F, 1982, p.. 31.)

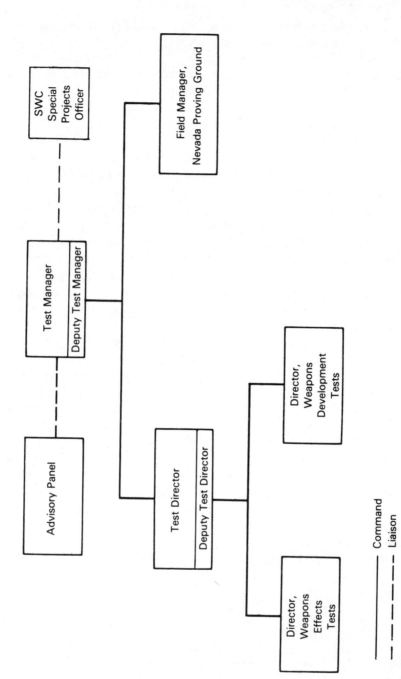

Figure 5. The Test Manager's interactions with his advisors out of the Santa Fe and Los Alamos, New Mexico Atomic Energy Commission Operations Office. (Source: Department of Defense, *Operation Buster-Jangle, U.S. Atmospheric Nuclear Weapons Test Personnel Review*, DNA-6023F, 1982, p. 32.)

was concerned that so large a detonation might produce serious shock in nearby communities or that it might cause severe fall-out or rain-out on more distant localities . . . and asked whether it was considered necessary to make a public announcement concerning the unusually high yield of the device to be tested. Mr. Graves . . . considered it unnecessary to go beyond the information contained in the draft press release on this subject . . . and pointed out that the greater power of this device would probably not be noticed by persons in nearby communities.[66]

At an AEC commissioners' meeting a year later, on October 5, 1954, a Los Alamos staffer, William F. Ogle, discussed the upcoming 1955 Teapot series, and noted that among the past difficulties that the test manager and the test director had had to deal with was the "pressure to complete on time." He also noted that the pressure was "permitted to affect the decision to fire."[67]

When test series were scheduled during the above-ground atomic testing period of the 1950s, the day-to-day planning of each shot was handled by the *test director*, a representative from Los Alamos. (Figure 6). While the test director was accountable for the daily planning up to the time of the detonation, the test manager had the overall responsibility for the shot and reported to the field manager in New Mexico. The test manager and test director had to adhere to a detonation time schedule developed jointly by the Armed Forces Special Weapons Project personnel and the AEC at Los Alamos.

This was the basic administrative and operational framework for the entire series of above-ground atomic tests. While the AEC commissioners in Washington endorsed the seven series of almost one hundred atomic test shots (ultimately approved by Presidents Truman and Eisenhower), they had no contact with the daily AEC policy interpretations of their field managers, test managers, and test directors, nor could they appreciate the daily pressures these AEC personnel felt in firing the atomic shots and immediately reporting the results to Los Alamos.

While the AEC commissioners and medical personnel had conveyed their concern about the safety of the AEC workers and the downwind civilians, the commissioners in Washington, D.C. never experienced the intense pressures that the local AEC operatives felt. The AEC's dilemma was magnified by the question of radiation safety for the civilian and military personnel on the site and for the citizens downwind.

The AEC and Radiation Safety

An inherent, fundamental conflict existed in the AEC from the moment of its creation in 1946. The federal agency was charged with the responsibility

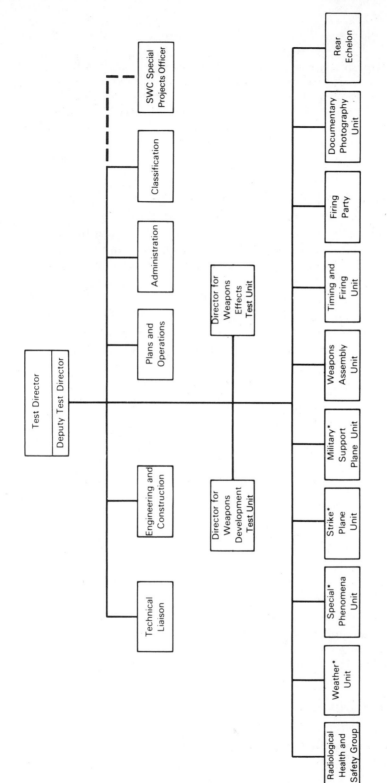

Figure 6. The Test Director's staff, his command, and his liaison contacts with the military personnel who were always directly involved in the testing program. The Test Director worked very closely with the Field Manager. (Source: Department of Defense, *Operation Buster-Jangle, U.S. Atmospheric Nuclear Weapons Test Personnel Review*, DNA-6023F, 1982, p. 35.)

of continuing, through its Division of Military Application, with the production of fissionable materials and the development of atomic weapons. The AEC was also given responsibility, through its Division of Biology and Medicine, for the health and safety of the participants in the testing program. Once the atomic detonations began, this extended to the people who lived downwind of the atomic blasts. The AEC commissioners *knew* there was an inherent conflict in their complex mission; at their December 13, 1950 meeting, when final approval was given to recommend to Truman the Las Vegas site, they stated that the problem of radiation safety was "the most central one in site selection."[68]

However, when the AEC had to choose between the necessity of atomic testing to maintain nuclear superiority over the Soviets and their serious concerns about radiation safety, conflicts were too often resolved in favor of the military's requirements. Eugene Zuckert, one of the AEC commissioners, said candidly, in 1954, that the nuclear testing program "resulted in a lack of balance between safety requirements and the requirements of the program. When cancer research studies came into conflict, the balance was apt to tip on the side of the military program."[69] As late as 1959, as testimony before the Joint Committee on Atomic Energy revealed, there was clear evidence of this continuing conflict. The Joint Committee concluded that the AEC's fallout research program "as a whole has apparently not received the high administrative level support it needs to give it the necessary impetus. Further improvements in this program, in its administration and its organization, are required, including adequate staffing in the AEC's Division of Biology and Medicine."[70]

The pathological and genetic dangers of ionizing radiation were known to scientists when the AEC began above-ground testing in the Nevada desert in January, 1951.[71] For decades articles in the medical literature had documented the causal relationship between radiation and several forms of cancer.[72] Dr. Charles L. Dunham, who during the 1950s testing period served as the Director of the AEC's Division of Biology and Medicine, acknowledged as much before the Joint Committee on Atomic Energy. In 1957, Dunham testified that, "since 1930, radiobiologists have developed an imposing amount of information concerning the effects of radiation on man."[73]

While knowledge existed about the correlation between radiation exposure and cancer, a number of critical questions remained in 1951 that challenged scientists doing research in this area. For example, was there a radiation *threshold* point below which a person would not suffer any long-term, harmful effects from radiation exposure? And, was there a cumulative effect on the health of those exposed to even the sub-threshold amounts, if such thresholds could be determined, such as the repeated, low-level exposure experienced by the downwinders?

In 1947, three years before the testing began at the Nevada Test Site, the

AEC started to formally explore these questions. According to Dr. Willard F. Libby, an AEC commissioner who testified before the Joint Committee in 1957, the AEC created the New York-based Health and Safety Laboratory in 1947 to review fallout data.[74] That same year, AEC Commissioner Lewis Strauss requested that the U.S. Air Force be given the responsibility for developing and implementing a radiation monitoring system to compensate for the lack of long-range systematic monitoring of radiation fallout in the atmosphere.[75] In addition, the Atomic Bomb Casualty Commission was organized in 1947, to conduct research on the short- and long-term impact of radiation exposure on the survivors of the atom bombs dropped on Hiroshima and Nagasaki.[76] During that same year, the AEC finally began studying the effects of radiation on laboratory animals (including fetal and young dogs and sheep) at Los Alamos and at the Argonne National Laboratory.[77]

In 1952, the AEC contracted with the RAND corporation to conduct Operation Sunshine, an examination of the ecological impact of radiation fallout, especially on soil and water. By the time the AEC's formal review ended in 1953, RAND had involved the assistance of AEC's National Laboratories at Argonne, Oak Ridge, and Hanford, and the help of university researchers at Columbia, Chicago, Rochester, and UCLA, the U.S. Department of Agriculture, and two other commercial laboratories in New York and Pittsburgh.[78] And, in 1959, President Eisenhower created the Federal Radiation Council to advise the President on radiation matters directly or indirectly affecting the health of persons in the vicinity of the testing program.[79]

However, the radiation threshold question continued to plague the AEC commissioners and their Division of Biology and Medicine staff. Dr. Donald S. Frederickson, M.D., Director, National Institutes of Health, Health, Education and Welfare, testified in 1979 that "by 1950, the International Commission on Radiological Protection and the National Committee for Radiation Protection concluded that there was no safe threshold."[80] In 1957, however, at the first public hearings ever held by the Joint Committee on Atomic Energy on nuclear test fallout, the testimony of the AEC medical experts minimized the danger of low-level exposure to radiation. Charles Dunham, the AEC Director of Biology and Medicine, wondered aloud whether the scientists who criticized the AEC's radiation safety reports were "so worried about the fallout as they are about the spread of nuclear war?"[81] Dunham assured the legislators that "there is no anticipated danger of fallout from the program."[82] However, Alvin Graves (test director in 1957), who had personally been exposed to radiation in the late 1940s, told them that radiation exposure meant that the danger of cancer or leukemia was greater. "Maybe the more likely is not very much more likely, but it is still more likely."[83]

At the 1959 Joint Committee hearings on the subject of radiation exposure, the legislators concluded that "a major unresolved question from 1957, i.e., is there a 'safe' minimum of radiation or 'threshold' below which there is no increase in the incidence of nongenetic conditions such as leukemia or cancer" was still unresolved.[84] The Committee Report concluded that the biological significance of low levels of radiation was still largely unknown. "No resolution was reached on whether or not a threshold level of radiation exposure exists below which effects such as leukemia and cancer do not result."[85] In 1962, another Joint Committee hearing report concluded, on the basis of AEC testimony, that the "threshold question is still unsettled within the scientific community."[86] Throughout the above-ground testing period, the AEC medical personnel insisted on the threshold concept even though other scientists believed that the AEC's safety standard was too liberal.

AEC medical advisors testified at congressional hearings in 1957 and 1959, and at the AEC commissioners' meetings, that accepting the more conservative safety figures *"would make it impossible to conduct operations at the test site without major changes in present procedures"* [my italics].[87] Dr. Gordon M. Dunning, Division of Biology and Medicine, in 1957 testimony before the Joint Committee, stated that the AEC scientists and technicians, "with moderate effort," could reduce even further the release of radioactive materials into the atmosphere.

> However, if we continue to reduce the fraction we are willing to release, we eventually reach a cost of control which makes the operation prohibitive. The dilemma is that we must weigh the degree of undesirability of radioactive fallout against the advantages which may be anticipated from activities which are inevitably accomplished by fallout.[88]

In a 1955 congressional hearing, AEC Chairman Lewis L. Strauss pointed out that "Soviet Russia possesses atomic weapons; there is no monopoly for the free world. . . . We have no alternative but to maintain our scientific and technical progress and maintain our strength at peak levels. The consequences of any other course would imperil our liberty, even our existence."[89] AEC commissioner Willard F. Libby, testifying at the 1957 Joint Committee hearings, put the situation in starker terms, contrasting the "very small and rigidly controlled risk" of radiation fallout with "the risk of annihilation."[90]

The AEC commissioners, however, expected that their agency's radiation standards would be properly implemented by the personnel at the site. In 1954, the AEC established a number of basic safety guidelines for the agency operatives at the site and elsewhere to deal with the dangers from exposure to radioactive fallout. Their obligations included (1) operating the program as safely as possible, (2) informing the public of radiation hazards and warning them in advance of the detonations, (3) reimbursing the public for any losses caused by the fallout, and (4) informing the public about the testing program activities.[91]

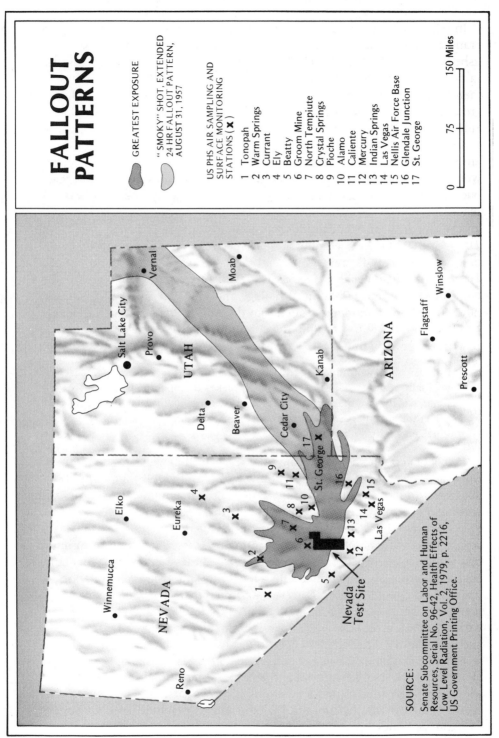

FALLOUT PATTERNS

GREATEST EXPOSURE

"SMOKY" SHOT, EXTENDED 24 HR FALLOUT PATTERN, AUGUST 31, 1957

US PHS AIR SAMPLING AND
SURFACE MONITORING
STATIONS (X)

1 Tonopah
2 Warm Springs
3 Currant
4 Ely
5 Beatty
6 Groom Mine
7 North Tempiute
8 Crystal Springs
9 Pioche
10 Alamo
11 Caliente
12 Mercury
13 Indian Springs
14 Las Vegas
15 Nellis Air Force Base
16 Glendale Junction
17 St. George

0   75   150 Miles

SOURCE:

Senate Subcommittee on Labor and Human
Resources, Serial No. 96-42, Health Effects of
Low Level Radiation, Vol. 2, 1979, p. 2216,
US Government Printing Office.

Source: Senate Subcommittee on Labor and Human Resources, Serial No. 96-42, Health Effects of Low Level Radiation, Vol. 2, 1979, p. 2216.

In 1953, the AEC established the Off-Site Monitoring Program in cooperation with the U.S. Public Health Service. The monitoring group was established under the direction and control of the Nevada Test Site field manager in order to provide the local AEC manager with information on the movement of radioactive clouds, the intensity of radiation in the air and on the ground, the duration of the radioactive fallout, and the size and shape of the fallout pattern.

The operating plan was a simple one which required seventeen stationary air sampling and surface monitoring stations and four mobile units that would be moved and concentrated in the region of maximum expected fallout. Public Health Service employees monitored the air and surface radiation readings in the following small locales in Nevada and Utah: St. George, Utah, and the Nevada communities of Mercury, Indian Springs, Las Vegas, Nellis Air Force Base, Glendale Junction, Alamo, Crystal Springs, Caliente, Pioche, Ely, Currant, Warm Springs, Tonopah, Beatty, Groom Mine, and North Tempiute.[92] The equipment used by the monitors was very rudimentary: film badges and portable monitoring devices for measuring surface radiation. The "primary air sampler was an Electrolux tank vacuum cleaner modified to collect materials" on the surface of a filter.[93]

The AEC's Off-Site Monitoring Program had the basic operational problems of frequent equipment breakdown, covering large areas of territory, interacting with the people in these small towns, and identifying and responding to "hot spots" (highly radioactive areas on the ground, surrounded by low activity). But *ground monitoring* of radioactive fallout was not begun until 1953, when the Public Health Service became involved in the monitoring program.[94]

This 1953 AEC/PHS ground monitoring program had its rough edges. During the 1953 tests, Nevada Test Site personnel stopped monitors from taking samples of local milk for possible radiation contamination. "Don't you do that," shouted an AEC on-site person to a monitor who was about to take milk samples. According to trial testimony, unknown, on-site personnel at the Nevada Test Site altered the radiation exposure readings taken by monitors in the local communities throughout the three-state area to ensure that the measurements fell within AEC standards.[95]

Frank Butrico was one of almost two dozen Public Health Service Radiation Safety Monitors working for the AEC in Nevada and Utah in 1953. He worked out of St. George, Utah at the time of one of the worst tests, "Dirty Harry," in the spring of 1953. "My instruments went off the scales," he testified in a 1982 trial, recounting that day. The radioactive cloud hung over St. George for over two hours, recalled Butrico. Fallout radiation levels peaked at a little less than 6 rads, well over even AEC standards. Although he himself burned his clothes and took a number of showers, Butrico was instructed by his AEC superiors at the Nevada Test Site headquarters only to give vague statements about radiation levels to inquiring residents in the small southwestern Utah town. When asked about the fallout, he told towns-people that radioactivity levels were "a little above normal [but] not in the range of being harmful."[96]

Butrico also recalled an AEC briefing for the monitors two days after Dirty Harry. "We are getting inquiries, some people have gotten sick—let's cool it - quiet it down," said an unnamed AEC official. "If we don't, there might be repercussions and they might curtail the program which, in the interest of national defense, we can't do."[97]

Finally, unknown AEC officials modified Butrico's memo to William John-son, the Off-Site Radiation Officer at the Nevada Test Site control center at Mercury, about the Dirty Harry fallout on St. George. Unnamed AEC staffers lowered Butrico's radiation readings and changed his report of the amount of time that elapsed before the AEC notified the St. George officials and school principals about the radioactive danger. The revised memo concluded with the statement, not written by Butrico, that the "effectiveness of the safety program was amazing." Butrico was outraged at this tampering. "I am disturbed," he said, in 1982 testimony before the federal district court, "that my name is over a report that contains statements unknown to me."[98]

What the AEC and the Public Health Service did not measure until 1957 was the *ingestion* of radioactive iodine into human thyroids. (Metabolized radioactive iodine goes directly to the thyroid gland, and is a basic cause of thyroid cancer.) Dr. Harold A. Knapp, a medical researcher with the Fall-Out Studies Branch of the Division of Biology and Medicine, AEC, from 1960 to 1963, found "no systematic collection of radiochemical analysis of milk and locally grown food products during the years of above-ground testing."[99]

As soon as the statistics became available, Knapp examined the AEC fig-ures about the power of detonations, and the general direction of radioactive fallout and weather conditions. He concluded that in the years of heavy test-ing (1951 to 1955), the radiation received by the thyroids of infants and young children who drank fresh milk was in the range of hundreds of rads. "The dose to an infant's thyroid from radioactive iodine in milk was 60 to 240 times the direct dose to the thyroid from external gamma radiation from fallout particles on the ground."[100]

It was not until 1962 that the AEC's medical research omissions from 1951 to 1957 were uncovered, more or less by accident, by Dr. Knapp. Dr. Knapp and others who researched radioactive iodine ingestion believed the medical impact on humans was devastating. In a 1965 report, Dr. Edward Weiss, another government medical researcher, concluded that a child's thyroid probably received a dose of between 200 and 1,200 rads. "This dos-age far exceeds the safety standards by a factor of 5 to 40 and is so high as to represent a serious potential cause of thyroid cancer."[101]

Drs. Knapp, Weiss, and other AEC medical researchers ran into unex-pected difficulty with the AEC bureaucracy in Washington during the 1950s and early 60s (when the 1950s atomic test radiation data was first analyzed by these scientists). When they requested permission to publish their findings about the possible adverse impact of low levels of radiation on the downwinders' health they were turned down by the AEC because, in the words of Dwight A. Ink, AEC Assistant General Manager, such studies would "pose potential problems to the AEC . . . (a) adverse public reaction; (b) law suits; and (c) jeopardizing the programs at the Nevada Test Site."[102]

Because of the AEC's fear of adverse publicity, a number of medical studies that focused on the dangerous effects of low-level radiation (written and completed in the early 1960s by Dr. Harold Knapp, AEC Fallout Studies Branch;[103] Dr. Edward Weiss, U.S. Public Health Service;[104] and Drs. John Gofman and Arthur Tamplin)[105] were not revealed to the public until Freedom of Information Act requests were made to the Department of Energy in 1978 and 1979. In addition to having their research curtailed, these medical researchers were occasionally confronted by legislators and AEC bureaucrats who were displeased with what they perceived to be efforts to derail the nuclear weapons development program.

Dr. Knapp recounted his conversation with a deputy director of the AEC's Division of Operational Safety. "When I told [him] how high the dosage levels were, the director had this pitch: 'Well, look, we've told these people all along that it's safe and we can't change our story now; we'll be in trouble.'"[106] Dr. Weiss was informed that his research was not going to be released, "because of potentially detrimental effects upon the government's nuclear weapons testing program."[107]

When Drs. Gofman and Tamplin testified before the Joint Committee, they were rebuked by an angry Committee Chairman, Chet Holifield (D-California): "What the hell do you guys think you are doing to the AEC's program? I don't give a damn who [you] are. You're going to have every little old lady in tennis shoes in this country up in arms against the AEC program. Listen, there have been lots of guys before you who tried to interfere with the AEC program. We got them and we'll get you."[108]

The medical research dilemma was clear: While the AEC standards called for radiation safety and AEC commissioners called for implementation of research programs by the Air Force, RAND, and their own medical personnel, research that threatened the continued viability of the testing program was hidden from peer and public review.

The AEC in the 1950s: Some Basic Dilemmas Emerge

The attitudes of the AEC commissioners and their staff in the 1950s clearly reflected the temper of the decade. America focused on maintaining nuclear superiority over the Soviets, and funds were provided (by the U.S. Government) for the expansion of facilities in order to produce the necessary fission materials, to develop tactical atomic weapons for use in the field, to develop the hydrogen bomb, and to stockpile the weapons at a rate far in excess of the Russians' capabilities.

While the AEC was concerned about the danger of low levels of radiation to citizens downwind of the Nevada Test Site, their concern was superseded by what seemed a more urgent, larger concern: the race against the Russians for nuclear superiority. Decisions were made, by the commissioners in Washington, D.C. and by the project and Nevada Test Site test managers, that enhanced the agency's capacity for research, development, testing, and production of atomic weapons for military purposes.

When downwinders grew concerned about the possible danger of fallout, the local Nevada Test Site managers dealt with residents' concerns by interpreting AEC regulations in their own fashion and used media public relations events to promote the atomic testing program. Films, lectures, and brochures were distributed by local AEC personnel to maintain "popular acceptance of the program."[109] Continually emphasizing the controversial threshold concept, AEC publicists claimed that radioactive fallout was not harmful; that it was no more dangerous than medical x-rays.[110]

The AEC field managers, concerned about the continuation of the atomic tests for national security purposes and fearful that adverse judicial opinions and negative public opinion would end the testing, acted in various, sometimes murky ways to maintain the Nevada testing program, as the AEC Commissioners in Washington, D.C. silently acquiesced.

Given the decentralized AEC organization that had been created by the AEC commissioners, and the major discretionary powers afforded the test site managers and directors at the site, it is evident that the AEC commissioners in Washington, D.C. were reluctant to make hard judgements about the tests. Their reluctance was also apparent in making decisions about the importance of safety precautions, and the dangers of radioactive fallout, and in ensuring that carefully developed safety plans would be correctly implemented for the downwind civilians and military personnel who might come in contact with the fallout.

As early as 1951, the AEC commissioners were perfectly content to have others take responsibility for radiation safety conditions at the Nevada Test Site. In a 1951 letter to the Military Liaison Committee, AEC Chairman Dean stated, "responsibility for troop compliance with this safety plan rests, of course with the [local] Exercise Director." The continuous refusal of the AEC commissioners to carefully monitor and oversee the activities of the Nevada Test Site operatives removed an important political counterweight to the continuing local pressures on the test site managers to "get" the shots off with no undue delay. While decentralization in any bureaucracy is generally valuable and necessary, the almost total abdication of safety responsibility by the AEC commissioners in Washington, D.C. meant that the local AEC technicians and the military officials were able to successfully exert enormous pressure on the test managers to approve the shot.

Administratively, beginning and ending the testing chain at the Nevada Test Site was appropriate; what was inappropriate and inexcusable was the AEC's inactivity despite the tremendous pressures the commissioners knew fell upon the shoulders of the local test directors. Although the minutes of AEC meetings reveal that the commissioners were aware of the stress placed upon the local directors (the commissioners debated these issues with the test directors), essentially all that came down from Washington, D.C. were messages from the AEC asking that "the concern of the Commission be conveyed to the test organization and that everything be done to avoid another fallout over St. George, Utah." This type of response was an ineffective counterbalance to the "pressure to complete on time" felt by the test manager.

Although the AEC commissioners, since 1947, had authorized formal studies by the AEC itself, by other public agencies, and by private research operations, on the impact of radiation on humans, plants, and animals, little was done in the 1950s to translate this knowledge into more tangible safety precautions. As late as the 1962 congressional hearings, AEC personnel were still defending the threshold concept. Further, using a benefit-cost analysis, the AEC commissioners themselves argued that it would be too costly to introduce additional safety precautions into the above-ground atomic testing program. One of their medical persons maintained that if "we continue to reduce the fraction [of radioactivity] we are willing to release, we eventually reach a cost of control which makes the operation prohibitive."

In addition, although the AEC commissioners continued to request that the necessary safety precautions be taken, there was never adequate funding. The tens of thousands of acres of land that lay to the east and south of the Nevada Test Site were monitored by seventeen units of off-site monitors, using inadequate equipment. Ground monitoring did not begin until 1953, and monitoring of radioactive deposits in the food chain did not begin until 1958, after the tests had ended. In 1962, a congressional committee concluded that radiation safety was horribly underfunded and strongly recommended that the Biological and Medical Division of the AEC receive additional funds for these needed safety studies, as well as for the implementation of more conservative safety standards. Finally, AEC staff in Washington, D.C., did all they could to minimize the impact of scientific studies that raised important questions about radioactive fallout and the health hazards to civilians.

During the 1951 to 1958 atomic testing, the events that took place at the Nevada Test Site, the AEC in Washington, D.C. and in the downwind communities, were fully known by the commissioners. However, the commissioners did not intervene to temper the zeal of the local operatives who either felt compelled, or were committed to the continuous development of new atomic weapons systems in order to maintain nuclear supremacy over the Soviets.

It was tragic that the local AEC managers never understood the religious and extremely patriotic Americans who lived and worked downwind of the Nevada Test Site. Had they known them, the local AEC personnel might have realized that by speaking squarely with them about the potential hazards of radiation and by informing them about the simple steps they could take to minimize the danger of radioactive fallout, the AEC not only could have avoided the negative public reaction they feared, they could have averted subsequent illness and death. Udall Stewart, a local Nevada rancher, said simply "If we had known about the fallout danger, we could have avoided it."[111]

3 The Downwinders' Response to the Atomic Testing, 1951–1958

The fallout settled on virtually uninhabited land.

AEC Publicity Statement

War is better than appeasing aggressors.

Iron County Record, 1951

I never saw a prettier sight; [the bomb blast] was like a letter from home or the firm handshake of someone you admire and trust.

Clint Mosher, Utah columnist, 1951

The testing is tragic and insane. But the play must go on.

Deseret News, 1953

I feel very much like a human guinea pig.   Mrs. Luana Buhler, 1957

I am frightened.   Mrs. Lois Grebbs, 1957

The year 1950 ushered in the bleak decade of Cold and Hot War. It began with the outbreak of hostilities in a place called Korea and ended with an ominous buildup of opposing forces, including some American military advisors, in a place called Vietnam. During 1950, these critical events occurred: *January*—Alger Hiss was convicted of perjury; *February*—Dr. Klaus Fuchs, a British physicist who participated in the World War II Manhattan project, confessed to passing atomic secrets to the Soviets; Senator Joseph McCarthy (R-Wisconsin), in a speech before a women's group in West Virginia, displayed his first "list" of 250 "known communists" who worked in the U.S. State Department; *March*—Fuchs was convicted; *May*—Harry Gold admitted passing atomic bomb documents to Fuchs; *June*—The Korean "police action" began; David Greenglass, one of Gold's accomplices, was arrested by the FBI; *July*—Julius Rosenberg was arrested and charged with passing secrets to the Russians; *August*—Rosenberg's wife, Ethel, was arrested and charged with espionage and passing secrets to the Russians; *September*—The Internal Security Act (McCarren Act) was passed over the veto of President Truman.[1] In summary, a somber and difficult year marked the beginning of a difficult decade.

In December of 1950, President Truman and the AEC commissioners made the tough policy decision to begin atmospheric testing of atomic bombs on and over the continental United States. The U.S. decided to conduct military testing of atomic devices in a relatively unpopulated area of the country in southern Nevada: the Nevada Test Site.[2]

The Downwinders

In 1950, the people who lived downwind of the Nevada Test Site, in Nevada, Arizona, and Utah, were predominantly Mormons. They were descendants of the first Mormon pioneers, who had arrived in the southwest from the eastern United States, the Scandanavian countries, and Great Britain in the mid-nineteenth century. The Mormon religion is central to an understanding of the downwinders' small-town life. For over ninety percent of the persons who lived in these communities, the Church of Jesus Christ of the Latter-Day Saints (the Mormon church) was the bedrock of their social, economic, political and religious existence.

The early Mormons, followers of Joseph Smith, came to their new religion "during a period of sharp social change and anxiety," which "brought into being a new culture as well as a new religion."[3] Joseph Smith, the son of an itinerant farmer, was the founder of their faith and self-proclaimed prophet of this new religion. He "had a dislike for the religious diversity that characterized nineteenth century America. Pluralism promoted controversy, and controversy encouraged social conflict, doubt, and disorientation."[4] Longing for order and unity rather than individualism, Smith's new religion emphasized cooperation rather than competition, and communal sharing rather than individualistic acquisitiveness. From its beginnings in western New York state in the 1830s, Mormonism "was a vital communitarian experiment,"[5] and a theocratic sect.

Smith and his followers created a religion that provided "a social refuge where priesthood leaders under divine inspiration would direct the political, social, economic, as well as the religious life of the community."[6] His new religion "sought to eliminate the economic and political reasons for social division by establishing a cooperative economic system . . . and by voting as a group."[7]

As the "prophet" for the new religion, Smith allegedly acted in God's name; having "received" visions from the Lord, he made decisions that affected every aspect of life for the members of the Mormon community. This decision-making pattern, the "principle of theocratic authority," quickly developed into the hierarchical structure that exists today (Figure 7). At the head of the "true church" is the three-man tribunal of the First Presidency. Beneath this is the Quorum of Twelve Apostles and then, progressing downward, are the High Councils, the leaders of the Stakes (each Stake is

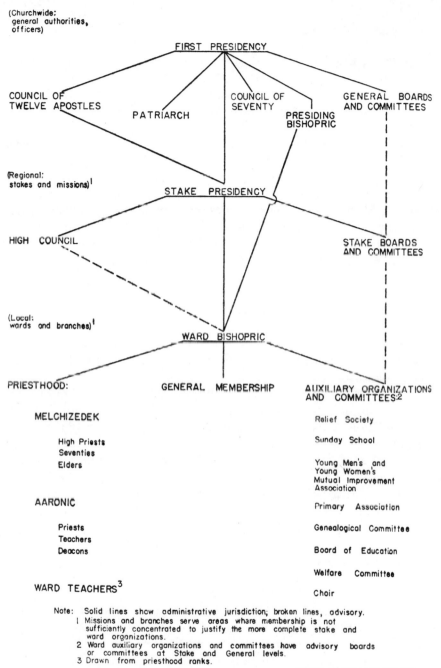

(Churchwide:
general authorities,
officers)

FIRST PRESIDENCY

COUNCIL OF
TWELVE APOSTLES

PATRIARCH

COUNCIL OF
SEVENTY

PRESIDING
BISHOPRIC

GENERAL BOARDS
AND COMMITTEES

(Regional:
stakes and missions)[1]

STAKE PRESIDENCY

HIGH COUNCIL

STAKE BOARDS
AND COMMITTEES

(Local:
wards and branches)[1]

WARD BISHOPRIC

PRIESTHOOD:

GENERAL MEMBERSHIP

AUXILIARY ORGANIZATIONS
AND COMMITTEES:[2]

MELCHIZEDEK

Relief Society

High Priests
Seventies
Elders

Sunday School

Young Men's and
Young Women's
Mutual Improvement
Association

AARONIC

Primary Association

Priests
Teachers
Deacons

Genealogical Committee

Board of Education

Welfare Committee

WARD TEACHERS[3]

Choir

Note: Solid lines show administrative jurisdiction; broken lines, advisory.
 1 Missions and branches serve areas where membership is not
   sufficiently concentrated to justify the more complete stake and
   ward organizations.
 2 Ward auxiliary organizations and committees have advisory boards
   or committees at Stake and General levels.
 3 Drawn from priesthood ranks.

CHART BY WILLIAM MULDER

Figure 7. The hierarchical structure of the Church of Jesus Christ of the Latter Day Saints. A great majority of the downwinders living in Nevada, Utah, and Arizona are members of this religious group, also called the Mormon Church. The towns dotting the map east of the Nevada Test Site had wards with their bishops; there were also some stakes located in the area. (Reprinted by permission of the University of Utah Press, from Lowry Nelson, *The Mormon Village*, copyright 1952, University of Utah Press.)

comprised of 4 to 8 Wards) and, at the lowest level, the Ward leaders. The Ward is the smallest unit of church organization and consists of approximately 150 to 300 families within a particular geographic area. Political, social, and religious judgements flow downward, from the President to Church members located in Wards across the globe.[8]

In the early 1800s, since their beliefs ran counter to the dominant Protestant religious thought of the period, the Mormons encountered bloody, religious persecution in New York, Ohio, Missouri, and Illinois. In 1847, after seventeen years of such conflict, and the murder of Joseph Smith and other early Mormons by virulent, anti-Mormon mobs, Brigham Young, a new leader and "prophet," led a small band of pioneers to the territory now known as Utah. Leaving Illinois for their land of "Zion" in the West, one of the pioneers, George Whitaker, said: "We are going to a better country where we can live in peace, free from mobs and strife, where we could worship God according to our conscience, none to molest or make us afraid."[9]

Utah was "the place" for Brigham Young and his followers, who crossed the Rocky Mountains and reached the magnificent Wasatch Front overlooking the Great Salt Lake Valley. The Mormons believed this western territory would provide them a safe haven from the religious bigotry they had encountered in the east.

But it was not to be. Soon after 1847, non-Mormon settlers, including Catholics, Jews, miners, and railroaders (all considered "gentiles" by the Latter-Day Saints) also arrived in the Salt Lake Valley after 1847, and for generations, there was tension between them. However, in March of 1895, after forty-odd years of clashing with the federal government, the U.S. Supreme Court, the U.S. Army, and "gentile" groups over the issues of polygamy, political power, theocracy, and capitalism, the people in the Utah Territory held a constitutional convention, and drafted ordinances prohibiting polygamous marriages and supporting womens' voting rights. The ordinances were approved by the voters of the Utah territory by a vote of 31,305 to 7,687 in November of 1895, and Utah entered the Union on January 4, 1896, as the forty-fifth state.[10]

In nineteenth-century, pioneer Utah, two distinct types of communities developed. "Gentile" communities, such as Fort Douglas, Park City, and Coalville, consisted of the individual homesteads and shops that sprang up around military, mining, or railroad operations. The Mormon farming village was a community built upon its inhabitants' communal and religious belief in the importance of family and church.[11] Unlike the isolated farmstead, which was "practically the universal system in America,"[12] the early Mormon settlements in Utah were essentially collective farms. Living in clusters and toiling together in the fields, these homogeneous people found happiness in their "social invention."[13] The village was established because of "a sense of the need to prepare a dwelling place for the Savior at His Second Coming."[14]

Examples of these early (1850s and 60s) Mormon settlements, each built at Brigham Young's order, are Cedar City, St. George, Kanab, Parowan, American Fork, Ephraim, Nephi, and Escalante, Utah, and the Nevada town of Bunkerville. Given their Mormon roots, it is not surprising that they all possessed certain "special characteristics": "The Church ramifies the social structure of the villages and is the dominant feature of it. The policies of the Church, therefore, affect behavior in the most remote community."[15]

One hundred years later, these Mormon towns retained the vestiges of their Mormon village origins. While there was a lessening of the formal Church dominance in the community life, these small, downwind towns continued to be, socially and morally, insular islands of Mormonism. An October 16, 1946 editorial in the *Iron County Record* (Cedar City, Utah) clearly illustrated the unbroken linkage between the early pioneers and their twentieth-century offspring:

> Utah's entire culture and economy have grown up with her towns. Each town [is] a thriving trade center in which the Church, a school, and a local weekly paper have been the guide and the inspiration. This was all inherent in Brigham Young's original colonizing plan. . . . It still shapes the living of more than half the population—ranchers, farmers, miners, artisans, and shopkeepers.[16]

During the 1940s and 50s, these towns exhibited all the characteristics of small-town, religious and social fraternities: their populations were almost 100 percent white and Mormon, with close-knit families (over 90 percent of the children lived with both parents). Just as in the 1850s, the "Mormon meeting houses play a prominent role in the community's social, as well as religious, life."[17] In 1950, the average family size in Utah was 8.83—the highest statewide figure in the nation. Utah had the highest birthrate in America, with almost 33 births per 1,000 population. The average Utah citizen finished high school, and 51.1% of the 18 to 21 age group, a record number nationally, were attending college in 1950 (up from the 1940 figure of 28.9%). Also, the average age of Utah's population in 1950 was 25.1 years, five years younger than the average age nationwide (30.2 years.)[18]

In these small towns, there was a small migration of younger people. Even with this exodus, however, the population of the towns had continued to grow, albeit very slowly, since their founding in the nineteenth century. (Washington County, for example, where St. George is located, had a population of 4,009 in 1890; in 1940, it was 9,270, and in 1950 it was 9,836. There was only a 6.1 percent increase in the St. George population statistics between 1940 and 1950. For comparison, urban Davis County to the north had a population increase of 95.6% over the same time period.[19]

Life in these small, post-war towns in Utah and Nevada was Church, Flag,

Mother, Apple Pie and Chevrolet: a religious, patriotic, ordered and simple life. One local newspaper revealed life as being, at times, too drab and simple. The *Iron County Record* commented on January 1, 1948, that "One of the greatest needs in Cedar City is a public dance and recreation hall, where wholesome recreation can be provided for the young people when school activities do not occupy their time. [This plan] has been met with indifference, and in some instances, with outright opposition, from community leaders."[20] The *Iron County Record* frequently gave front page coverage to the tragic consequences of the community's not having a place for young people to go: terrible automobile accidents involving young people who were driving to neighboring small towns that had such facilities.[21]

The townspeople were employed in agriculture (especially sheep and cattle farming), construction, tourism (Zion and Bryce Canyon National Parks are just two of the scenic attractions in southwestern Utah), small manufacturing (no "smokestack" industries), retail trades, and professional occupations. Women played a minor role in the labor force in this part of the country during the late 1940s and 50s for two basic reasons: (1) The religious-social importance of child rearing. There were fewer two-paycheck families in Utah than in most other states because Mormon women were expected, according to Church doctrine, to raise as many children as they could. (2) Light industry jobs that were suitable for women were scarce.[22]

A contemporary economist, writing in 1956, noted that a new industry had mushroomed in Utah, Colorado and New Mexico. In 1948, the "Uranium boom hit Utah and a modern gold rush started. . . . There is one buyer, the Atomic Energy Commission, which is prepared to buy all ore containing a minimum of uranium at a specified price."[23] Given the Atomic Energy Commission's push in the direction of nuclear power, uranium (once a low grade by-product of vanadium milling) "suddenly became the valuable major product leading to the atomic bomb."[24] In 1955, there were over 2,000 men working the uranium mines in Monticello and Marysville, Utah; in 1948, less than 200 men had been mining uranium.[25]

"Typical of things to experience in [these] small agricultural towns," wrote a contemporary geographer, ". . . are the odor of new-mown hay; the smell of dust, barnyards and rain; and the sound of wind, running irrigation water, frogs, birds, and crickets."[26] The townsfolk who lived downwind of the Nevada Test Site also joined rose clubs and had vegetable gardens. "There are very few people in town who don't have a garden. It's a tradition."[27] They raised their own vegetables and produced their own meat and milk. (Popular features in the local weekly publications were the food canning section and the recipe column.)

The adults in these small towns were thrifty and friendly; they disliked "big-city" life and were content to live in these clean, relaxed and secure small towns that were, to them, extended families. "I like the people. . . .

There are really no rich or poor. They are willing to help. They work hard and are just good people."[28]

In the post-World War II days, townfolk kept up with current events with their local weekly newspapers, such as the *Iron County Record* (Cedar City) and the *Washington County News* (St. George), and by tuning in to a small radio station beamed out of Cedar City. These small, local media outlets provided the people in the small towns with local news about births, deaths (including, evidently, the sometimes weekly occurrence of tragic automobile deaths in the region), socials, weddings, and above all, the religious news of the area.

The Mormon Church also developed a strong media program that blanketed small-town Utah and the surrounding states. WKSL, a powerful, Mormon Church-owned radio station (the most watt-powerful station in the West) beamed national news to these small towns from Salt Lake City a few hundred miles to the north. The *Deseret News*, owned and operated by the Church, was published in Salt Lake City (with local news bureaus throughout the state and region). It was and is the largest statewide, daily newspaper in Utah. The Church also owned and operated KSL-TV, a commercial television station, and KBYU-TV (and FM), a public broadcast television and radio station transmitting its signals across the west from the campus of Brigham Young University in Provo, Utah.

Social patterns in these small towns stemmed from the Church's position on major ethical issues. The Church's standards included very conservative positions on sex, especially homosexuality. Mormons were also directed to abstain from cigarettes, alcohol, caffeine and other stimulants. Church doctrine has espoused a very proscribed role for women in social life (the Church opposed the Equal Rights Amendment) and, until 1978, its religious doctrines were racially discriminatory. (This discrimination was repudiated by Spencer W. Kimball, the Church's "prophet," who received a "vision" on June 9, 1978 that led to an overnight change in Mormon policy regarding the role of blacks in the Church, a change that was called the "greatest moment in the modern church."[29])

Two contemporary political observers have stated that today "in Utah the 1950s American ideal is still alive and thriving."[30] That ideal, which involved total personal commitment to Church, family, and flag, was reflected in the extreme pride and fierce patriotism of the townspeople in southern Utah and Nevada. Their attitudes about the Cold War and the need for nuclear testing were clear and uniform: Communism must be stopped, and it was the responsibility of the American government and people to see that communists were licked.

Shortly after the Korean War began, in June of 1950, the local newsweekly in Cedar City ran a story on August 3 about plans for a home guard in Utah. "An MP company will be formed in Cedar City as part of a home guard in defense of invasion against air invasion or communist-sponsored revolts in Utah," read the article. It concluded by calling upon "disabled or over-age vets to volunteer for the home guard."[31] And an editorial that appeared in the *Iron County Record* on January 4, 1951, underlined the courage and determination (along with an early expression of the "better dead than Red" attitude) of the rural, religious Westerners, and declared that:

> the people are not terrified of another war and the possibility of atom bombs being used against us. War is better than appeasing aggressors. The people are ahead of the government in their thinking; the people are perfectly willing to face the facts; . . . the people will rise to this crisis [Korean War], bitter as it may be, as they have risen to all others.[32]

A political cartoon on the front page of the March 1, 1951 *Washington County News* typified the local citizens' patriotism and pride in American nuclear military might. Entitled, "Can This Be 'The Thing?'," it depicted Joseph Stalin and Mao Tse-Tung being knocked off their feet by the "Booms" of the atomic devices being tested at the Nevada Atomic Testing Grounds (Figure 8).[33] It implicitly illustrated the local people's commitment to the testing program and their pride in having an important defense program right in their own back yards. They believed they were a part of the fight against godless communism and strongly supported the continental atomic testing program taking place just a few hundred miles away. Even more important, their faith and trust in their government would not allow them to even consider the possibility that the government would ever endanger their health.

After a series of nuclear test shots had deposited radioactive dust on St. George and other downwind towns in May of 1953, the *Iron County Record* broached the question of "Atomic Winds—Fall-Out,"[34] in a lead editorial. While noting his concern about the possible health hazard of the unexpected fallout over the local towns and speculating about the impact of this national news story on the Cedar City tourist trade, the editor relied on the words of "Dr. Gordon Dunning, representative of the Atomic Energy Commission Division of Biology and Medicine." Dunning's final words to a Cedar City reporter, faithfully quoted in the editorial, were comforting ones: "It has been shown that the levels of radiation produced outside the test control area were in no way harmful to humans, animals or crops."[35]

In addition to "the sense of patriotism" there was a "sense of civic obligation imparted into the lives and value systems"[36] of people living downwind of the Nevada Test Site: As Professor James Mayfield has observed, "Clearly, the Mormon cultural influence in Utah generates commitment."[37] For the townspeople of southwestern Utah and neighboring Nevada and Arizona, the act of voting was, and is, seen as a fundamental civic responsibility. Given the Church's power and influence, in this area of the country

CAN THIS BE "THE THING?"

Figure 8. This 1951 cartoon, from a St. George, Utah weekly paper, illustrates the attitude of the citizens residing in these small southwestern towns. The atomic test blasts at the Nevada Test Site are shown literally knocking Stalin and Mao Tse-tung off their feet. (Reprinted by permission of The *Washington County News,* March 1, 1951.)

"national election day is among the most popular community rituals. . . . Mormon theology emphasizes the moral obligation that every Latter Day Saint member has in fulfilling his civic duty."[38]

Election statistics bear out the fact that since 1920, Utah has been continually among the top eight states in voter turnout. Since the presidential election of 1940, Utah has never been ranked lower than the fourth highest state in voter turnout percentages. (It was fourth highest only once, in the 1944 election. In all other presidential elections Utah has been among the top three states in voter turnout.[39]) For example, in 1940, 80% of Utahns voted in the presidential election compared with 60% nationally (second highest state turnout in America); in 1944, 76% voted, compared with 56% nationally (fourth highest turnout); in 1948, 75%, compared with 52% nationally (highest turnout); and in 1952, 81% voted, compared with 62% nationally (third highest turnout).[40] The Mormon Church's admonition to

participate in this critically important civic function has been strongly
adhered to by the people who lived downwind of the Nevada Test Site. Rural
voters in the area tended to vote in greater numbers than their urban cous-
ins. They have "more involvement in local affairs as well as a comparatively
greater political optimism and awareness of the individual utility of the political process."[41]

Because of the average Utahn's appreciation of the New Deal Democratic
programs, Democrats dominated the political scene during World War II.
The shift to the Republican Party in the state of Utah occurred during the
1940s. It evolved out of the hard work of the Mormon Church leadership
who had "discovered," in the 1930s, a "new found devotion to laissez-faire
economics."[42] According to Gottlieb and Wiley, greatly perturbed by the
"evil of the dole," and by the fact that the federal government was providing
the kind of help that could be better provided through Mormon self-help
programs,[43] Mormon Church leaders began the arduous task (because the
lay Mormons liked New Deal programs so much) of "slowly weaning away"
Utah voters from the Democratic Party.[44]

In the aftermath of the Depression years and the War, and entering a
period of economic prosperity,[45] Utah shifted its voting support away from
the Democrats. Because of the new-found prosperity,[46] combined with traditional, conservative social values and the encouragement of the Church,
the state's voters moved into the Republican Party column. In addition,
communism had become, by 1950, the major public policy issue in the state.
In the U.S. senatorial elections that year, Mormon Church officials "joined
the Republican Party in a campaign that they hoped would rid the state of
the New Deal."[47]

The 1950 senatorial campaign was an extremely dirty one, typical of
many ugly congressional campaigns during the early stage of the Cold War.
It was a "graphic example of McCarthyism in action."[48] The *Deseret News*,
the Church-owned and operated paper, supported the Republican challenger, Wallace F. Bennett. The newspaper, which reflected the Church's
ideological commitment to Republicans, linked Bennett with "faith and freedom," and labeled U.S. Senator Elbert Thomas, the Democratic incumbent,
an "atheistic, communist New Dealer."[49] The paper further accused Thomas
of presiding at communist meetings and sponsoring leftist organizations, and
called him a puppet of communism and the radical unionists. Battered by
this barrage of untruths, the New Deal incumbent was narrowly defeated.
His Republican challenger, Bennett, never retracted these allegations.[50]

In this obdurate fashion, the Republican Party era was ushered in. After
two decades of chipping away at the credibility of the Democratic Party,
using the powers and machinery of the Mormon Church, the Church leaders
had effectively shattered the strength of the New Deal party in Utah and
surrounding states.[51]

In the rural counties in southwest Utah, the shift to the Republican Party had taken place somewhat earlier than the statewide shift. For example, Iron County, Utah (Cedar City) has voted Republican in *every* presidential election since 1932, with the exception of the 1936 election. (A slight majority of the votes from the County in that election went to Franklin D. Roosevelt.)[52] Washington County, Utah (St. George) has voted for the Republican presidential ticket in every election since 1948.[53]

Contemporary news accounts and voting statistics clearly demonstrate that politicians residing in the small towns downwind of the Nevada Test Site—conservative, patriotic, religiously loyal followers of the Mormon Church leadership in Salt Lake City—found that the Republican Party more faithfully reflected their religious, economic and social values.[54] Politically, Utah became one of the most conservative states in the Union in the 1940s, and it is still a bastion of conservative strength in American politics.

The Downwinders' Reactions to the Nuclear Test Series

Seven major test series were conducted during the dangerous decade of the 1950s. In the years 1951, 1952, 1953, 1955, 1957, and 1958, almost one hundred atmospheric detonations of atomic devices occurred, one hundred brilliant flashes burned into the darkness, and one hundred dangerous radioactive cloud patterns drifted away from ground zero. Due to changing wind currents, poor weather forecasting, or operator-carelessness, almost thirty percent of the radioactive debris from the atomic bombs moved over these communal, largely Mormon towns that lay to the east of the Nevada Test Site.

The townsfolk, over 100,000 men, women and children, were scattered in the states of Arizona, Nevada and Utah. Although they were shocked and frightened, at times, about the atomic blasts, these people believed that the testing was vitally important for America's national defense. And while they wished the testing could be done elsewhere, they managed to conduct their lives—so they thought—with little interference from the nearby atomic explosions. The *Deseret News* editorial echoed their views: if testing were necessary for national security and if appropriate precautions were taken by the Atomic Energy Commission, then "we'll go along."

At Meeting 504 of the Atomic Energy Commission, on December 12, 1950, the Commission approved a staff recommendation that the Las Vegas site be used as the continental proving grounds for atomic weapons. "The Las Vegas location," the report stated, "most nearly satisfies all of the established criteria for a continental test site."[55]

The loyal, religious, and vigorously anti-communist citizens living downwind of the Nevada Test Site saw the continental atmospheric nuclear testing

WARNING

January 11, 1951

From this day forward the U. S. Atomic Energy Commission has been authorized to use part of the Las Vegas Bombing and Gunnery Range for test work necessary to the atomic weapons development program.

Test activities will include experimental nuclear detonations for the development of atomic bombs — so-called "A-Bombs" — carried out under controlled conditions.

Tests will be conducted on a routine basis for an indefinite period.

NO PUBLIC ANNOUNCEMENT OF THE TIME OF ANY TEST WILL BE MADE

Unauthorized persons who pass inside the limits of the Las Vegas Bombing and Gunnery Range may be subject to injury from or as a result of the AEC test activities.

Health and safety authorities have determined that no danger from or as a result of AEC test activities may be expected outside the limits of the Las Vegas Bombing and Gunnery Range. All necessary precautions, including radiological surveys and patrolling of the surrounding territory, will be undertaken to insure that safety conditions are maintained.

Full security restrictions of the Atomic Energy Act will apply to the work in this area.

RALPH P. JOHNSON, Project Manager
Las Vegas Project Office
U. S. Atomic Energy Commission

Figure 9. Handbill distributed by the Atomic Energy Commission in 1951 just before the very first series of tests took place at the Nevada Test Site. (Source: Department of Defense, *Ranger Tests,* p. 20.)

program as a necessary governmental policy, and welcomed it. The down-winders were told by the Atomic Energy Commission, in 1955, that as cit-izens, they were "active participants in the Nation's atomic test program."[56] However, from the first tests conducted by the Atomic Energy Commission and the Department of Defense in 1951 through the "Hardtack" nuclear test series in 1958, the downwinders in Utah and Nevada became aware of the term, "radioactivity." Slowly over the decade, the concern of the townsfolk about the danger of radioactive fallout grew. And, repeatedly, the Atomic Energy Commission assured them that they were safe.

The testing of atomic devices on the continental United States began in the winter of 1951. On January 11, 1951, Atomic Energy Commission employees distributed a handbill, framed in black, warning residents to keep away from the Nevada Proving Grounds because of test activities that would be taking place at the site "from this day forward" (Figure 9). In addition to the ominous looking flyer, the local weekly newspapers in the southern counties in Nevada and Utah ran stories that instructed citizens to keep out of the area. The *Washington County News*, for example, on January 25, 1951, ran a story that stated "Officials of the Las Vegas office of the Atomic Energy Commission today warned all unauthorized persons to stay off the Las Vegas bombing and gunnery range." The Atomic Energy Commission officials were concerned about "careless, curious, and hard-to-reach persons" such as prospectors, live stock tenders, and private fliers.[57] Shortly after these warnings were issued by the Atomic Energy Commission, the initial tests were conducted.

1951

Ranger Series (January 25–February 6)

On January 27, 1951, for the first time since the Alamogordo blast in 1946, an atomic bomb was detonated in the United States. A *Deseret News* headline blared: "A-Bomb Exploded in Nevada Tests: Second U.S. Blast Rocks 250-Mile Area."[58] This was the first of the "periodic blasts" referred to by the Atomic Energy Commission in its January 11th flyer. The blast was seen and heard in northern Utah (Tremonton); Salt Lake City; San Diego, San Fran-cisco, Fresno, Needles, and Los Angeles, California; and by residents in the local downwind communities.

The detonation startled many persons early on that Saturday morning, including a mailtruck driver in Orderville, Utah (who thought the Russians had dropped an A-bomb on Los Angeles); a farmer on his farm, five miles from Beaver, Utah (who said that when the flash occurred, his cows "put their tails in the air and ran helter skelter"); and a truck driver, south of Las

Vegas, who said that "the flash blinded me for a second or two and gave me quite a scare. I have seen the Northern lights often—but this explosion made them look silly."[59] (A Salt Lake woman called the *Deseret News* to report "'a very bright blast.' Two minutes later she called again. 'I was wrong,' she said. "I was looking east and what I saw was the sun coming up."')[60]

James Bundy, the Mormon Ward columnist for the *Washingon County News*, didn't "care too much about the thrill. It makes me feel that war is right in our back yard, and I don't like war any place, any time, with anyone."[61] However, Nora Lyman, in her February 1, 1951 *Observation* column that appeared in the *Washington County News*, was more relaxed about the Nevada Test Site activities of the previous weekend. After talking about the abundance of sunshine in Utah and the new furniture on sale at Pickett's, she recounted her feelings about that first blast:

> Early last Saturday morning, the city was agog with excited reports of having heard and felt an earthquake, . . . it was not until a noon radio news broadcast that we learned of an A-bomb explosion near Las Vegas. . . . In St. George, windows rattled pretty well all over the city and many persons saw the light quite in advance of hearing the noise. . . . Atomic physicists call their trick of producing low order nuclear fission burst without provoking the atomic blast into a full-scale, death-dealing bomb 'tickling the dragon's tail.' The blasts, both Saturday and Sunday mornings, were pretty elated tickling, I'm thinking. I could not help thinking of the destruction, confusion, and annihilation that would have resulted had that been a regular A-bomb explosion. The AEC assured everyone that the test was held under the full security restrictions and that there had been no reports of 'radio-logical hazards.'"[62]

If people were concerned about possible danger to their health from the blasts, the Atomic Energy Commission took steps to assure them that there was nothing to worry about. "A-Bomb Blast Said Safe: AEC Officials Say Hazard Not Possible," headlined the *Deseret News*.[63] Dr. John Z. Bowers, a University of Utah medical professor and Dean of the College of Medicine, stated that the Atomic Energy Commission was "monitoring with great exactness the progress of the bomb cloud to see if there is any deposition of radioactive material," and declared, "I can't see any possibility of trouble . . . We have more than adequate assurance of no contamination to human life."[64]

By the third blast, the *Deseret News* could report that local residents were becoming so accustomed to the repeated tests that "the sheriff's office [in Las Vegas] reported it had received not a single telephone call a half hour after this third explosion." And, in the gambling casinos, "dealers continued to flick cards to '21' players . . . without doffing a green eyeshade."[65]

The fourth test, a "Tremendous A-Blast Seen in Four States," was headlined by the *Deseret News* on February 2, 1951. The United Press reported that "for the first time since the tests began, a dense cloud over the mountain-ringed proving ground was visible from Las Vegas. An Atomic Energy Commission spokesman said in answer to queries that it was a dust cloud."[66] After this test, the police headquarters was inundated with telephone calls from frightened citizens and "burglar alarms were set off 'all over town.'"[67] By this time, reports had been published in the *Deseret News* that "A-Tests Spread Radioactive Snow Over East."[68] The United Press report indicated that "atomic snow" had fallen in Rochester, New York; Cincinnati, Ohio; Chicago, Illinois; and in Schenectady and Upton, New York. Atomic Energy officials, however, quickly discounted the danger of radioactivity to the residents of these cities far from the Nevada Test Site.[69] After the fifth and final A-bomb sunflower had detonated, people were once again "calm" and telephone calls to police and pressrooms ended.[70]

By and large, the local people reacted positively to this first series of atmospheric atomic tests. Congressmen did not receive letters from constituents complaining about the shattering blasts. Atomic Energy Commission Chairman Gordon Dean reported that the Agency had received some mail but no complaints. Instead, he offered the following apocryphal story: "We even heard of one fellow who's planning to sue the government for damages sustained when one of our blasts shook a pair of dice on a Las Vegas gaming table from a 'sure seven' to 'snake eyes'. He claims it busted him."[71]

While there was some concern about the possible loss of tourist traffic to the southwestern towns surrounding the test site ("Vegans Fear Effect of Atoms' Blasts: Divorce, Gambling Trade May Slump, Residents Say," read a headline in the *Deseret News*)[72], and a developing interest in "Home A-Bomb Shelters,"[73] the local news media's reaction to the tests was very positive— and patriotic. The *Washington County News* reported to its St. George readers that the Atomic Energy Commission had "Concluded A-Bomb Tests at Las Vegas Site" and that "'use of the site,' said Carroll Tyler, NTS field manager, 'has saved manpower, materials, money and, above all, invaluable time in the national atomic energy development program.'"[74]

The *Deseret News*, on January 28, 1951, ran an editorial entitled "Spectacular Atomic Explosions Mean Progress in Defense, No Cause For Panic."[75] The editorial writer for the Mormon-owned paper spoke with pride about America's new tactical atomic weapons and concluded with the following words: "The people of Las Vegas, it is reported, were startled but not panicky. This is as it should be. There is no reason for panic in the fact that the Army is experimenting on improving its tools."[76] Another *Deseret News* editorial, February 4, 1951, dealt with the possible structural damage an A-bomb might cause, and concluded that there was little danger:

> Such danger is slight, and probably negligible. . . . The zone of destruction of the A-bombs is strictly limited, and most of the destructive force has gone through the air, not through the ground. And even the force of an atomic bomb

explosion is small in comparison with the energy released during a good-sized earthquake.[77]

Articles and opinion columns in the newspapers read by the downwinders emphasized "Atomic Research To Aid Mankind,"[78] "A-Bombs: The Greatest National Asset,"[79] and "Las Vegas A-Bomb Tests Of Interest To Russia."[80] The latter column reflected a popular theme that struck a very responsive chord in the minds and hearts of the small-town downwinders: patriotism and anti-communism.

Harry Ferguson wrote in the *Deseret News* of January 30, 1951, that while "it's a long way from Las Vegas to Moscow, there isn't any doubt that the blast that rocked Nevada over the weekend also was felt in Russia."[81] Ferguson saw the testing program at the Nevada Test Site as a "shield" that protected America and Western Europe from a "Red invasion."[82] The Russians had been following the testing with fear and interest and believed that America was maintaining its superiority over the communists through the nuclear tests. "Stalin won't have any trouble figuring out what was going on at Las Vegas. There were tests. You hold tests when when you develop new things, new techniques, and new weapons. In other words, we have advanced considerably beyond the A-bomb that fell on Hiroshima and brought Japan to her knees in World War II."[83]

Aside from citizens' complaints about broken windows and the complaints made to the Atomic Energy Commission by a few unlucky craps players, the initial local response to the 1951 testing was that it was a necessary part of our nation's strategy to combat the Russians and other threatening communist nations. As Clint Mosher wrote, after observing the last blast: "I never saw a prettier sight; it was like a letter from home or the firm handshake of someone you admire and trust."[84]

Buster-Jangle (October 22–November 29)

The Buster-Jangle tests, in the fall of 1951, also did not cause any major excitement. The downwinders were getting used to the blasts and the aftershocks. Two of the bombs blasts were "terrific" and were felt in St. George.[85] Nora Lyman, in her *Observation* column in the *Washington County News*, told a cute story about two young travelers who were given a free night's lodging in the local jail—only to have the sheriff go hunting without remembering to let the couple out. She then recounted her personal experience with the big blast:

> Did you feel the tremors or hear the blast of the atomic bomb Tuesday morning? I believe it was the greatest vibration we have had in this area. I was in the kitchen preparing breakfast when I heard a sound like that of an explosion of dynamite, and the first wild thought which ran through my mind was that our butane gas tanks were exploding. At that instant the Boss said, 'Someone is

rattling our back door. See who it is.' Then I realized that it must be an atomic bomb explosion.[86]

The *Iron County Record* was silent about the fall tests of 1951, with the exception of some syndicated stories that commented that the "U.S. Needs Civil Defense,"[87] and a small filler about how the Atomic Energy Commission had worked out arrangements with the Civil Aeronautics Administration regarding airplane overflights in the Nevada Test Site vicinity. While the *Deseret News* reduced the scope of its coverage of Nevada Test Site activities, the newspapers' commitment to the atomic testing program was unchanged. In a "Good News" editorial on November 4, 1951, the good news was that "the U.S. Army ran through a series of atomic tests which were 'most successful.' The significant point was that it was indicated that this country now has perfected atomic weapons which can be used against troops in the field—perhaps atomic artillery shells or a compact aerial bomb."[88]

By the end of the first year of atmospheric tests, the downwind townspeople had concluded that the atomic tests were a vital part of America's defense. Little concern was voiced about "radio-logical" health hazards (as Nora Lyman worded it in her chatty column in the St. George weekly). The bomb was seen as "a letter from home."

1952

Tumbler-Snapper (April 1–June 5)

By the spring of 1952, the atomic testing at the Nevada Test Site received less coverage in the St. George and Cedar City newspapers than had the two earlier test series. Nora Lyman did not chat about the test that woke her up and the *Iron County Record* had no stories about the Tumbler-Snapper tests. The *Washington County News* contained one small article that announced a "Nuclear Test Detonation At Nevada Proving Site."[89] Even the *Deseret News* began to treat the nuclear testing as business as usual. General stories were published about the military testing of tactical weapons "with atomic warheads as small as a grenade," weather effects and misfires. In addition, a series of articles were written by reporter John R. Talmage, on the "Terrifying A-bomb, The Symbol Of Our Age".[90] But these did not have the intensity of the earlier coverage and Talmage's articles focused on the story of the atom, isotopes, chain reactions and other easy-to-read points on nuclear physics.[91] The testing program had become just another story for the people downwind and for the news editors. A *Deseret News* editorial entitled "The Best of All Reasons for Peace," noted that the atom bomb was becoming familiar but that "makes it no less terrible."[92]

However, one incident marred the placid *Deseret News* coverage of the

1952 tests. In retrospect, it was a watershed event. Lyle Jepson, a Salt Lake City radio technician, won a $10 prize from the *Deseret News* for calling the paper to report that there were dangerously high levels of radioactivity over Salt Lake City.[93] On May 8, 1952, the *Deseret News*, thanks to Jepson's tip, ran a story headlined "S.L. A-Dust Declared Harmless."[94] The paper stated that the reporter contacted Atomic Energy Commission officials in Las Vegas about the higher than average levels of radioactivity over Salt Lake City. The Atomic Energy Commission spokesperson responded by saying that the radioactive cloud had been charted by the testing agency, and since there was no danger to the residents, no one was informed. "The AEC spokesman said he could not 'emphasize too strongly the absence of dangerous effects from the radioactivity in the air over Salt Lake City.'"[95]

The following day, however, a *Deseret News* editorial, "Atoms in the Dust," expressed great concern about the disturbing radiation incident:

> All of us have known for the past six years that we are in the midst of the Atomic Age. But until Wednesday, Salt Lakers did not realize just how literally the atomic age was in their own midst. Wednesday's sudden rise in the level of atomic radiation in Salt Lake and other Utah communities provided moments of startled incredulity, grave concern and later heartfelt relief when it was learned that the level of radioactivity was not dangerous . . . although, the level of radiation [was] enormously above 'background' levels. . . . Intermountain residents will now be more keenly aware of potential deadliness in the air each time a nuclear device is exploded on the nearby Nevada Proving Grounds. They want to be sure that the AEC is completely aware of its tremendous responsibility in assuring the safety of the entire area before an atomic explosion is set off. . . . Precautions must be redoubled, never relaxed.[96]

However, no further incidents were recorded by citizens and the 1952 tests ended with no further editorials or letters to the editor. The Atomic Energy Commission did not schedule tests at the Nevada Test Site until the following spring—one year later and enough time for cooling off. During the interim, people and news editors turned to other issues.

1953

Upshot-Knothole (March 17–June 4)

In the spring of 1953 eleven "nuclear events" involving "realistic nuclear field exercises" with military personnel took place on the Nevada desert.[97] For the people living downwind, it was an extremely worrisome time because of the number of "dirty bombs," i.e., detonations that deposited nuclear radioactive fallout on their small towns.

On March 5, 1953 local residents read in their newspapers that the

Atomic Energy Commission was going to begin testing in the very near future. The Cedar City paper quoted the AEC spokesperson as stating that "experience . . . has proven that nuclear testing can be carried out with adequate assurance of safety."[98] On the 17th of March, the test series began at the Nevada Test Site. As a part of its recently developed public relations program, the Atomic Energy Commission had invited a small number of women to see the atomic blasts. Mrs. Rae Ashton of Vernal, Utah, and national president of the American Legion's Womens' Auxiliary, gave her impressions to *Deseret News* readers: "I really believe that some day science could destroy the whole world with a bomb like this. . . . Maybe it's just a woman's logic. But there must be some way to live without using something like this on objects with warmer blood than store dummies."[99] Klien Rollo, another eyewitness from Cedar City, reported on the first page of the *Iron County Record* of her "good fortune to be invited" to the atomic detonation: "The light was so bright that it was impossible to recognize colors, if that is possible."[100] She noted, with implicit concern, that "Governor Val Peterson, head of the National Federal Civil Defense Administration, was flabbergasted by the amount of dust and debris present even at two miles distant from ground zero."[101]

After the initial shot in the Upshot-Knothole series, the *Deseret News* editorialized on "Man and Doom in the Desert." It was a solemn essay that called the testing of both defensive and the new tactical, offensive atomic weapons "tragic and insane," and continued, "But the play must go on. So long as we live in an atomic world, we must and will continue to learn more about this power and how to survive it. . . . It must go on until—God willing—man turns away once and for all time from the evil power driving civilization toward the grave."[102]

Eight days later, the *Deseret News* ran the first of what was to become a stream of stories over the next few years on the possible dangers of radioactive fallout in the clouds that drifted over the downwinders after the detonations: "A-Cloud Dangers In S.L. Studied; University Nuclear Expert Finds High Rays Concentration." The comments of Dr. Lyle E. Borst (a former director of the Atomic Energy Commission's Brookhaven National Laboratory) directly referred to the dilemma that confronted the downwind residents during the 1950s: *were radiation exposures cumulative or did the radiation effects wear off after a short period of time?* The Atomic Energy Commission's (non-threshold) position was that there was no cumulative effect and that "small doses received a long time apart are not cumulative." Borst, however, disagreed and was quoted in the news story:

> When an atom bomb is dropped "there will be people somewhere on the fringe for whom the difference between life and death will not be the amount of radiation from the bomb, but the amount of radiation they have previously

absorbed." . . . Dr. Borst said he had not allowed his children to play outside [after a detonation] and had required them to wash their hands and bathe more frequently than usual. "I don't believe in taking chances," he said. "I would no more let my children be exposed to small amounts of radiation unnecessarily than I would let them take small doses of arsenic. The only time a person should expose himself to radiation is when there is a very definite and good reason."[103]

In an editorial, "The Safest Way," the following day, the *Deseret News* picked up the major theme underlined by Borst: the uneasiness that some Utahns were starting to feel about the impact of radioactive fallout following the atomic tests. While it was "comforting" to the downwinders that the Atomic Energy Commission has "given assurance that the level of radiation experienced so far cannot possibly be harmful to human beings," Borst's message was of "increasing concern."

> Extra concern is felt for the residents of southwestern Utah, roughly from Cedar City south, who fall within the 200-mile circle around the proving grounds which is covered by a security blackout. *The public is never told just what levels of radiation are reached in this area*, except that the AEC assures us that they have been 'well within the limits of safety'—whatever those are. *The 'limits' are the precise subject of the debate* [my italics].[104]

The editorial then offered a practical solution to the dilemma, reflecting the willingness of the downwind citizens to continue their participation in the efforts to maintain the security of the land they loved.

> When an atomic cloud passes over any area, [the] civil defense organization could immediately call a full dress rehearsal for the public, just as though a cloud of dangerous intensity were threatening the region as a result of enemy bombing. The benefits would be two-fold: people would learn what to do in case of actual enemy attack and at the same time they would be giving themselves the fullest possible protection from whatever real or potential dangers the nuclear radiation may carry.[105]

The Atomic Energy Commission did not act on this editorial suggestion, even though it had taken note of the Borst's concerns. According to the minutes of Meeting 866, on May 22, 1953, Henry D. Smythe, one of the AEC Commissioners,

> read excerpts of a letter from Mr. Lyle Borst, formerly employed on the reactor project at Brookhaven Laboratory and now at the University of Utah, in which Mr. Borst expressed concern about the radioactive fallout resulting from the UPSHOT-KNOTHOLE series of tests. [The letter] *was further evidence of the concern in the minds of the population in the vicinity of the test site* [my italics].[106]

At the same meeting, another AEC commissioner, Eugene M. Zuckert, said that the fallout problems made him fear "testing a device at Nevada considerably larger than any previously fired there. *A serious psychological problem has arisen*, and the Atomic Energy Commission must be prepared to study an alternate to holding future tests at the Nevada Test Site" [my italics].[107]

In its formal *public* response (which appeared in the *Deseret News*), however, the Atomic Energy Commission had Dr. John C. Bugher, director of the Atomic Energy Commission's Division of Biology and Medicine, issue a statement discounting Borst's concerns about levels of radioactive fallout. "Danger Claim Disputed: A-Cloud Safe, Utah Told," read the *Deseret News* headlines. Borst refused to comment on the AEC's statement, telling the *Desert News* he did not want to argue with anyone "in the columns of a newspaper." "However," he said, "my children are going to continue to stay indoors whenever an atomic cloud passes."[108]

Upshot-Knot test stories, which appeared in the press after the March, 1953 incidents, continued to emphasize the radioactivity danger. "Two Miners Found Radioactive—110 Miles From A-Blast,"[109] "Excess Radioactivity Delays 8th A-Blast,"[110] "A-Radiation Ruled Out in Mystery Death,"[111] were some of the headlines in the *Deseret News* during the test series. After a mid-May shot allegedly sickened some Orderville miners in southern Utah, the Atomic Energy Commission sent experts "to Check Sick Utahns."[112] However, the paper reported, "*the AEC emphasized that the mission was to reassure area residents that they could not possibly have been harmed by radioactive fallout*"[113] [my italics]. The following day, May 21, 1953, a *Deseret News* headline read: "AEC Denies Blast Caused Utah Illness."[114] The AEC, it was reported, "flatly denied that radiation could have caused illness reported by a group of southern Utah miners."

There was confusion, however, among the Atomic Energy Commission investigators. The *Desert News* reported that some of them told the mine operator, Sam Ellett, that "wind currents could have been responsible for unusually heavy fallout. But Frank Waters, assistant information officer for the Nevada Proving Ground, said that Ellett has 'misquoted our people.'"[115] That day's *Deseret News* editorial (May 21) reminded the Atomic Energy Commission, if the operators were reading the paper, that "human life is too precious to be risked in experimentation and guess-work."[116]

The local weekly newspapers in southern Utah, however, were still much less critical of the Atomic Energy Commission's Nevada atomic testing program. A May 21, 1953 editorial in the *Iron County Record* assured its readers that everything was still safe for "humans, animals and crops."[117] Most local stories were essentially reprints of Atomic Energy Commission press releases: "Atomic Artillery Shell Proves Successful; Another May Be Set Off,"[118] "AEC Explains 'A' Blast Recording Organization."[119]

Shortly after the mid-May shot, Angus Bulloch and other sheep ranchers lost thousands of head of sheep, in an area close to the Nevada Test Site, "by some mysterious malady."[120] Nevada health authorities and the AEC issued a tentative report stating that "malnutrition was probably the cause of the loss rather than suspected radio-activity caused by the recent bombings at Frenchman's Flat in Nevada."[121] On June 25, 1953, the *Iron County Record* ran a short story headlined: "'Malnutrition' Explains Sheep Mystery Disease."[122] (This was the first coverage of this incident, which led to protracted litigation in federal courts in 1956 and from 1982 to 1984.[123] See Appendix A for the full story of the sheep death litigation.)

While most of the nuclear testing stories and editorials that appeared in the local newpapers were noncontroversial in nature, Cedar City's *Iron County Record* ran a controversial feature story in early May. The story, written by a University of Utah student, Ralph J. Hafen, was extremely critical of the Atomic Energy Commission's atomic testing program. Hafen felt "morally obligated" to tell the Cedar City downwinders of the

> possible irreparable damage that may have occurred or may in the future occur [as a] consequence of the continuing series of nuclear explosions in Nevada. . . . The public is day after day exposed to radiation that may very well be injurious. Your health, your children's health, and the health of generations yet unborn, are at stake."[124]

The student raised serious, jolting questions about radiation in general, and specifically discussed questions about inhalation of plutonium, eye cataracts, recent heavy fallout in Cedar City and St. George, and radiation and the mutation rate. The Atomic Energy Commission had not answered these questions, he wrote, and there is controversy surrounding them. However, Hafen concluded, "the gravity of the problem is such that we cannot afford to allow [the tests to continue] while the argument is being settled. The burden of proof is on the AEC. They are morally obligated to clear up the above questions before continuing with their tests in Nevada."[125] By the end of the 1953 test series, the radiation issue was clearly identified by the media, especially by the *Deseret News*, as an issue of major concern for the downwind citizens.

At this point in the testing program at the Nevada Test Site, 1953, a "folklore" began to emerge out of the strange incidents that all seemed to be related to the atomic testing.[126] Strange burns afflicted the downwinders and their horses, and mysterious illnesses killed cattle and sheep. After many of the atomic tests, geiger counters owned by downwinders who were former part-time prospectors in search of uranium lodes registered off the scale. These readings suggested that radiation levels were high, yet official Atomic Energy Commission reports indicated that the levels of radioactivity were insignificant. In some of its news releases, the Atomic Energy Commission attempted to discount the geiger counter phenomenon. Because the downwinders and the media had no knowledge of the exact amount of radiation deposited on the people, cattle, vegetables, and desert sands of the downwind region, and desert ranchers and prospectors circulated stories about geiger counter readings of over 50 to 100 rads,[127] speculation and fear abounded.

While most of the people of southern Utah and Nevada remained patriotic, some downwinders' patriotism was tempered by this growing concern that the Atomic Energy Commission was not being truthful to them about the exact amounts of radioactivity that had fallen or about the real danger the fallout posed for them, their children, and their children's children.

During 1953, Atomic Energy Commission personnel in Washington, D.C. became uneasy about the downwinders' growing fear regarding the atomic tests, as revealed in the minutes of the May and June meetings in 1953 (post-Upshot-Knothole). The Atomic Energy Commission was concerned about the viability of the Nevada Test Site as a continental test site, due to the rumblings about fallout health hazards in Utah and Nevada, and the Atomic Energy Commission's "serious public relations problems which had developed from the fallout incidents accompanying the test series and the importance of presenting immediately to the public the full facts concerning the reported incidents."[128] One of the participants at a June 10, 1953 meeting claimed that

> the people in the vicinity of the Nevada Proving Grounds no longer had faith in the AEC, and there was discussion of the bases for this feeling and of the importance of choosing, for an objective presentation of the AEC 'case,' men who would enjoy the full confidence of the public.[129]

At this time, local Atomic Energy Commission operatives initiated a serious public relations effort that included "a public education program."[130]

At a July 7, 1953 meeting, Lewis L. Strauss, Chairman of the Atomic Energy Commission, commented that one of his chief staffers

> had expressed to him his concern that the Commission might have underestimated the seriousness of the fallout problem. Mr. Smythe and Mr. Zuckert said that there had been no disposition on the part of the Commission to think that the fallout problem was not a most serious one."[131]

The emotional climate in 1953 was one of growing concern of not only prospectors, ranchers, *Deseret News* editorial writers, and other downwinders, but also of the Atomic Energy Commission, university professors, and doctors, regarding the radiation problem. While there were strong calls for the Atomic Energy Commission to exercise extreme care and openness in implementing the testing program, there was no serious outcry for a halt to the atomic testing in the Nevada desert.

The Atomic Energy Commission formulated new operating plans, in 1953, that would deal more carefully with the fallout program, and AEC Commissioners began to question the need for atmospheric testing. At Meeting 962, on February 17, 1954, Commissioner Smythe questioned "the desirability of avoiding open shots in the future if it is at all possible to do so."[132] Others at the meeting discussed the appropriateness of continuing to use the Nevada Test Site itself.[133]

However, the relationship between the Atomic Energy Commission and the downwinders continued on a downward path. The 1953 tests had clearly unnerved these religious, trusting people, and the Atomic Energy Commission was unable to reassure them. Incredibly, the Atomic Energy Commission frequently appeared to be as concerned about the fallout dangers as were the downwinders.

1955

Teapot (February 18–May 15)

In October of 1954, a few months before the commencement of Operation Teapot, the Atomic Energy Commission had a Teapot briefing in Washington, D.C., conducted by William E. Oagle, a scientist from the Atomic Energy Commission's Los Alamos Scientific Laboratory (LASL). His message to the harried commissioners was that the "successful completion of a test series, *from the standpoint of safety*, would be determined by the nature of fallout control problems" [my italics].[134] Oagle assured the Atomic Energy Commission leaders that "detailed, last-minute weather information" would resolve the difficulty.[135] With the aid of the latest weather data, "formulae had been developed which could be applied to TEAPOT to identify the shots which were likely to be serious from the fallout aspect. The serious shots could then be fired during weather conditions which could not produce serious fallout in the inhabited areas."[136] Apparently relieved, AEC commissioners gave final authority to begin the 1955 atmospheric tests at the Nevada Test Site on February 11, 1955. The commissioners approved the new radiological safety criteria and hoped that the radiological dilemma would go away. It didn't.

During the 1955 test series, the *Deseret News* continued to plead for extreme care and, for the first time, toward the end of the series, for the removal of the atmospheric test program from the Nevada Test Site. The local newspapers in St. George and Cedar City began to voice serious concern about the radiation fallout and its impact on the health of the downwinders, and the Atomic Energy Commission began to hear from senators living in the testing region.

In St. George, Utah, and other downwind communities, various Atomic Energy Commission and U.S. Public Health Service officials met with local citizens to discuss public health and safety. This type of public relations operation was duly noted in the *Washington County News*, on February 17,

1955—one day before the Teapot tests began.[137] "A conference regarding public health and safety aspects of the nuclear tests at the Nevada Test Site beginning this week was held Saturday afternoon," the weekly newspaper reported. In attendance were Dr. Gordon Dunning of the Atomic Energy Commission and scientists from Los Alamos Scientific Laboratory and the U.S. Public Health Service. It was a combined effort to assure the audience that "representatives of the AEC are stationed in all communities adjacent to the test area, and are willing and anxious to be of service to the local people."[138]

In January, 1955, a month before the Teapot series began, James E. Reeves, the Nevada Test Site Test Manager, addressed a message to the "people who live near the Nevada Test Site":

> You are in a very real sense active participants in the Nation's atomic test program. You have been close observers of tests which have contributed greatly to building the defenses of our own country and of the free world. . . . Some of you have been inconvenienced by our test operations. At times, some of you have been exposed to potential risk from flash, blast or fallout. You have accepted the inconvenience or the risk without fuss, without alarm and without panic. Your cooperation has helped achieve an unusual record of safety. . . . I want you to know that each shot is justified by national or international security need and that none will be fired unless there is adequate assurance of public safety. We are grateful for your continued cooperation and your understanding.[139]

This warm and friendly rhetoric appeared in an Atomic Energy Commission pamphlet given to school children and their parents, as part of the AEC's new efforts to gain the support of the local communities downwind of the test site. And, for a little while, the AEC's strategy worked.

A headline appeared in the *Iron County Record*, on February 24, 1955, that read: "Atomic Tests Start in Nevada; No Measurable Fallout." The article stated, "Residents of southern Utah have been assured through the various monitoring stations in the area that no radiation fallout from either of the blasts has been detected."[140] Residents who read the newspaper articles with toned-down messages about the nuclear tests were not too concerned with the initial Teapot shots.

However, at Atomic Energy Commission Meeting 1062, held in Washington, D.C., on February 23, 1955, Atomic Energy Commission Chairman Strauss read a letter from Senator Clinton P. Anderson (D-New Mexico), Chairman of the Joint Atomic Energy Committee, to his fellow commissioners. In the letter, the southwestern legislator asked whether the Nevada Test Site could be

> utilized effectively and economically . . . for anything other than very small yield devices. . . . I would appreciate receiving from the AEC an evaluation of the NTS under the present criteria for test operations. Your report should consider

whether only very small yield devices should be tested there, leaving all substantial shots for the Pacific where they can be precisely scheduled. . . . I am
raising the question of whether we can use the NTS efficiently for anything other
than the test of very small yield devices.[141]

The AEC commissioners were quite concerned because it was "quite a big
change in Anderson's attitude."[142] Anderson and the rest of the Joint Committee had worked closely with the Atomic Energy Commission, and the
commissioners felt that a change in the relationship at that time would further exacerbate problems at the Nevada Test Site. The commissioners were
concerned because they did not want to shift the testing program to the
Pacific. "I think this will set the weapons program back a lot to go to the
Pacific," said Commissioner W.F. Libby.[143] Strauss, however, wanted to take
the two large kiloton (as yet unfired) Teapot devices, load them on a ship,
"and go out to Eniwetok and put them on a raft and set them off."[144] Commissioner Libby hoped "that the furor will die down as we go through the
series."[145]

Realizing that the fallout dilemma would continue, Libby said that "people have got to learn to live with the facts of life, and part of the facts of life
are fallout." Strauss responded sarcastically, saying, "It is certainly all right,
they say, if you don't live next door to it."[146] Commissioner, Thomas E. Murray, added a few words: "We must not let anything interfere with this series
of tests—nothing."[147]

A final, critically important observation was made by Chairman Strauss
about the communities downwind of the test site. Strauss noted that distinct
fallout patterns had developed over the previous three years of atomic testing. "The fallout apparently always plasters St. George," said the Atomic
Energy Commission Chairman. "*Of course*," he added, "*they* [the AEC
operatives] *really never paid much attention to that before*" [my italics].[148]

Meanwhile, less than a week later, the *Washington County News* ran a small
story on March 3, 1955 indicating that the St. George mayor and other city
officers "feel that there is no danger as the fallout is not great enough to
cause danger. *Men who are connected with the Atomic Energy Commission*
explained to them the fallout and compared the degree of this fallout with
an x-ray, so they felt there was no cause for worry" [my italics].[149]

The *Deseret News*, however, continued in its editorials to criticize the testing. In a March 8, 1955 editorial entitled "For the Future of the Race," the
Mormon Church newspaper urged America to overcome our "knowledge of
of our ignorance of the possible long-range effects of radiation. . . . It is still
inconceivable that men should destroy themselves and their posterity with
any kind of knowledge of what they are doing."[150]

A day later, the Atomic Energy Commission detonated a large nuclear
device, and a banner headline appeared in the *Iron County Record* on March

10, 1955: "Fallout? 'Not Enough to Worry About,' Says AEC." The reporter noted in the article that "in all likelihood the cloud was from the atomic explosion," and it "dispersed directly over Cedar City. . . . Now the question of fallout has become real and some people in the area with geiger counters have expressed themselves that they believed an extensive fallout occurred in this area."[151]

Even Nora Lyman's "Observations" column in the *Washington County News* began to lay out the two sides in the fallout controversy. After the next blast, she wrote, "Atomic bomb explosions on the Nevada flats have caused a considerable amount of fear among the people who live within a radius of 100 miles or so, including, of course, St. George. [Government] doctors and officials sent into the areas say that radiation is about one-twentieth of that experienced in an x-ray, although geiger counters and scintillators jumped around quite a bit."[152] (In this column of May 5, 1955, Nora recounted that her trip as an observer at the Nevada Test Site was "thrilling"—even though the Atomic Energy Commission cancelled the shot at the last moment.)[153]

Throughout the Teapot tests, the Atomic Energy Commission's public relations program continued at the local level. On April 14, 1955, the *Washington County News* noted that the "Chamber of Commerce Hears Discussion, Sees A-Bomb Fallout Picture." Sandwiched in between a report on the Easter egg hunt and the building of a western street for a Hollywood motion picture to be filmed in the town, it was reported that the St. George Chamber of Commerce members were given the "premiere showing of the Atomic Energy Commission's 'fallout' picture of which parts were filmed in St. George."[154]

For the first time since the atmospheric tests began at the Nevada Test Site, stories appeared in the local papers about the *Bulloch* litigation proceedings instituted against the federal government for human injuries and animal deaths. An article in the *Iron County Record*, March 3, 1955, noted that "Local Sheep Raisers File Suit With Govt For Loss." The story reported that the sheep ranchers, Bulloch, and others, were

> convinced that atomic fallout from the experiments at Yucca Flats in Nevada two years ago was the direct cause of heavy loss of sheep herds grazing in the adjacent areas. . . . Not at all satisfied with the efforts made by the AEC and the government to resolve the problem, local livestock people have filed suit against the government to collect damages totalling almost $200,000.[155]

(While the ranchers were unsuccessful in their 1956 litigation against the federal government, recent events have enabled them to reopen their trial in Federal District Court in 1982, almost three decades later! See Appendix A.)

Then, on May 4, 1955, the *Deseret News* reported on its front page,

"Nevada Woman Claims Radiation Caused Cancer." The story indicated that Mrs. Martha Sheahan, owner of mine property at Groom, Nevada, had filed suit for $75,000 in damages against the Atomic Energy Commission, "charging radiation exposure from the atomic test site in 1953 caused a cancer on the left side of her face, below the eye."[156]

At the conclusion of the Teapot tests, the *Deseret News* printed a hard-hitting editorial entitled simply, "Don't Hurry Back." While complimenting the Atomic Energy Commission for its commitment to its own fallout guidelines, keeping the faith, and thereby holding up the Nevada Test Site tests at a cost of $50,000 per day until conditions were right, the newspaper voiced the feelings of many downwinders with these observations:

> It would be hard to find any Utahns who would go into mourning if the AEC decides to abandon its Yucca Flats nuclear tests permanently. . . . We Utahns have put up with sleeping next to a tiger decently enough—as have the even more directly concerned Nevadans. If the tests are necessary to national defense and as long as careful precautions are taken, we'll go along. But if the AEC decides to conduct subsequent tests in the ocean or Antarctic or somewhere else, they'll find no complaints here.[157]

The Teapot test series ended on May 15, 1955. The Atomic Energy Commission soon decided to continue the atmospheric testing at the Nevada Test Site but wait until 1957. While there were renewed public relations efforts by the Atomic Energy Commission, including films, lectures, and personal observations of the nuclear blasts by local reporters, the radiation fallout issue lingered. Fallout was a fact of life, as the AEC Commissioner had noted.

1957

Plumbob (April 24–October 7)

During the Plumbob series, with twenty-four tests, the Atomic Energy Commission had to confront tough congressional reviews of radiation and its effects on humans, a hostile media, and a large community of downwinders, in three states, who were growing progressively more disenchanted with the Atomic Energy Commission and the Nevada Test Site tests.

While Plumbob, the sixth atmospheric test series conducted by the Atomic Energy Commission, did not lead to dirty detonations, as the 1953 Upshot-Knothole tests had, the radiation issue continued to be the major controversy in the newspapers, in Atomic Energy Commission sessions, and, for the first time, before the Joint Atomic Energy Committee of the U.S. Congress. As soon as the Plumbob series began, the public was made aware, through the congressional hearings that began in May, 1957, of the "una-voidability" of health hazards due to radiation exposure from atomic fallout. (Increased radiation hazards were an "inseparable" result of increased military testing, said a witness who appeared before Senator Clinton Anderson's committee in late May, 1957.)[158]

Also, for the first time, the Atomic Energy Commission received criticism for whitewashing the danger radioactivity posed to humans. The agency's attitude toward the danger of radioactivity was sarcastically labelled the "body-in-the-morgue" approach, i.e., the Atomic Energy Commission grew concerned and took the radiation problem seriously only when the danger was "readily noticeable."[159] A Herblock political cartoon which appeared in the Deseret News on July 1, 1957 (Figure 10) was a trenchant commentary on the Atomic Energy Commission's efforts to minimize the impact of publicity about dirty bombs and fallout.

In Utah and Nevada, however, the scenario remained the same. Days sometimes began with bright white, early-morning atomic explosions, followed by brownish-purple mushroom clouds, then pinkish clouds which drifted over the small towns and ranches; and children played in the grey radioactive dust as they would in snow. The Atomic Energy Commission stepped up its public relations gimmicks: an "Atoms for Peace" Mobile Unit visited Cedar City and other downwind towns in the summer of 1957,[160] an "Atomic film" was shown to the local Rotary Clubs and Chambers of Commerce, and Atomic Energy Commission booklets were distributed to school children (in 1955 and again in 1957) entitled, *Atomic Tests in Nevada* (see Appendix C). Bomb shelters were advertised; local civil defense officials attended nuclear fallout programs sponsored by the Atomic Energy Commission;[161] and stories appeared such as "Fallout Scare Strikes Tiny High Sierra Town,"[162] while the Atomic Energy Commission issued solemn pronouncements declaring that the amount of exposure that had resulted from the tests was "no worse than having a tooth x-rayed."[163]

It was also in 1957 that the *Bulloch* litigation began in federal district court,[164] peace protestors opposed to the Plumbob tests[165] demonstrated at the Nevada Test Site in August, and newspaper editors received a steady influx of letters from citizens who were concerned about the radiation issue.

Given the townspeople's heightened sensitivity to the radioactivity issue, the Atomic Energy Commission had to address the downwinders' use of geiger counters. To combat the stories about geiger counters and their high radiation readings, the Atomic Energy Commission issued statements such as, "A geiger counter can go completely off-scale in fallout which is far from hazardous"[166] and "Many persons in Utah, Nevada and Arizona, and nearby California have geiger counters these days. We can expect many reports that

Figure 10. From *Herblock's Special for Today* (Simon & Schuster, 1958).

'geiger counters were going crazy here today.' Reports like this may worry people unnecessarily. Don't let them bother you."[167]

As usual, *Deseret News* editorials tried to manage a balance between the patriotic mood (the tests are necessary for national security) and the angry mood (get the bomb blasts out of the desert and set the damn things off in the ocean—somewhere, anywhere, but the Nevada Test Site). Typical Atomic Energy Commission press releases circulated in the *Washington County News* and *Iron County Record*, which announced the Plumbob tests and invariably published reports (next to stories like "LDS Seminary Undertakes Book Cataloging") and indicated that "AEC monitors found no trace of radio-active fallout from the test."[168]

In addition to troublesome congressional hearings on the health effects of radiation and prospectors reports of high radiation readings on their geiger counters, the Atomic Energy Commission had to deal with another unexpected headache during the 1957 test period. Dr. Albert Schweitzer, a world-famous, medical humanist, called for an end to atmospheric testing because of the incalculable dangers of radioactive fallout to unborn generations of children. One AEC commissioner, Willard F. Libby, responded to Schweitzer's concern and conceded that *"radioactivity may exceed safety limits in some places if peacetime tests are continued at the present rate"* [my italics]. But this problem could be avoided or overcome, said Libby, by having people fertilize the soil or import their milk![169]

On April 27, 1957, the *Deseret News* printed an editorial in a fit of American nuclear jingoism in an editorial entitled "When To Stop Atom Tests." This was the Church's response to Schweitzer's plea for a nuclear test moratorium. "Can we, should we, end our tests?" queried the paper, concluding that the testing should not be halted until the Communists agreed to a moratorium. Regarding the threat of radioactivity, the editorial alluded to Libby and the Atomic Energy Commission who "point out" that "a person suffers eight to forty times more radioactivity from a wrist watch than comes from radioactive fallout." In this mood, the editorial concluded that the United States had two worldly missions: "to teach this story of radioactivity to the rest of the world and to work for world peace so that atomic tests can be stopped safely."[170] The Church's position was best summed up in an August 15, 1957 political cartoon (Figure 11) in which a stern Uncle Sam told a peace protester to convince Khrushchev and the Kremlin that A-bomb testing should be stopped. (Protestors were in Nevada at the time, protesting the Nevada Test Site testing.)

The congressional hearings were of critical concern and importance to the Atomic Energy Commission. The radiation dangers that had previously been hidden by the Atomic Energy Commission were now under the glare of television cameras. Nationally syndicated columns were being written by observers such as Eric Severeid. His essay, "The Fallout Debate Goes On,"[171] enlarged the discussion to the national and international public. The hearings provided the Atomic Energy Commission with an opportunity to put their best face forward on the critical issue of radiation and health hazards. In the late spring of 1957, the agency did just that, and soothed the congressional critics of continental atmospheric testing.[172]

Locally, the Letters to the Editor column began to carry the concerns of the downwinders. "Are Intermountain Folks Guinea Pigs?" asked Delome Billings, on June 25, 1957. His letter pointed out that radiation danger is cumulative, that life spans are shortened from radiation exposure, and that meat was contaminated by the radioactivity. "What kind of folly is it that

We've Always Been Convinced

Figure 11. This 1957 Shoemaker cartoon from the *Deseret News* underscores the government view that, while the United States would like to halt the testing of atomic devices, it could not until the Russians stopped testing their atomic devices. (Courtesy of Vaughan Shoemaker.)

we allow ourselves to be subjected to this danger repeatedly? Perhaps to the Atomic Energy Commission authorities, the people of this intermountain section are considered expendable and convenient guinea pig material, but I protest it emphatically."[173] On July 3, 1957, Luana Buhler's letter was printed. In it, she shared Billings' views:

> I, too, feel very much like a human guinea pig. . . . There must be countless areas in the western portion of the world where these tests could be conducted without endangering the health of populated area. . . . We may be expendable,

but I don't agree; and if the men who are conducting these tests have that opinion, they need replacing.[174]

Lois Grebbs, a young mother of "six small children" wrote from Sandy, Utah that she was "frightened about what might happen to those little bodies . . . who have been exposed to seven times the amount of radiation fallout considered safe." She also commented about the "unborn" children: "Atomic tests are not worth sacrificing our children and our children's children for. Let's stop them now."[175] On September 17, 1957, Mrs. D. Billings wrote to complain about the 1957 Nevada tests that were soon to end. She pleaded with the paper to "do what you can to obtain a cancellation of the proposed A-bomb test scheduled for September 18, 1957 at the Nevada Test Site." People would be showered with radioactive particles, she claimed. "For the health and welfare of all the men, women and children in the Western United States; to maintain unpolluted the animal and vegetable life; to leave undisturbed the equilibrium of the good earth, please do whatever it is possible for you to do."[176] These letters were plaintive pleas for action to end the tests at the Nevada Test Site, and although they did not lead to action, the letters were part of a citizens' rebellion against the bland assurances of the Atomic Energy Commission. By this time, most downwinders felt that the Atomic Energy Commission had lost its credibility.

The commissioners in Washington, D.C. and the AEC and Los Alamos Scientific Laboratory operatives at the test site were unable to end the nagging concerns of the downwinders. Although the federal agency spent millions of dollars on "technicolor" films, pamphlets, mobile units, displays, lectures, and meetings, the Atomic Energy Commission could not silence the geiger counters or the fears of the downwinders.

One day after the tests ended, the *Deseret News* editorialized about "Operation Plumbob."[177] While "the whole operation, it must be admitted, has been eclipsed, from the viewpoint of public excitement, by Russia's spectacular launching of the first satellite, . . . the nuclear experiments in Nevada have proved of value in a nuclear struggle."[178] In a statement that reflected the Church's position on nuclear war and international affairs,[179] the editorial concluded with a theological perspective:

> The only real way to stop nuclear warfare or preparation against such an awful thing, is to inculcate into the hearts of all men of all nations, the spirit of brotherhood, understanding and love. When this is done, bomb blasts will no longer be fired. All who love peace should pray that this happy day will come speedily.[180]

Needless to say, this was not the response the letter writers were hoping for from the *Deseret News*. Shortly thereafter, the Atomic Energy Commission approved the seventh series of atomic tests, Hardtack, Phase II.

1958

Hardtack, Phase II (September 9–October 31)

By the time of the 1958 tests, the United States was committed to testing the hydrogen bomb. Phase I of the 1958 Hardtack test series was conducted in the Bikini Island area in the Pacific Ocean testing grounds. Hardtack, Phase II, at the Nevada Test Site was to be the last test series scheduled by the United States government before an official moratorium on aboveground atomic testing.

On August 22, 1958, President Eisenhower announced that the United States, joined by Great Britain, was ready "to withhold testing of atomic and hydrogen warheads for one year to facilitate a broad system of international controls. . . . The proposal has one major condition: that in the next one-year period, starting October 31, 1958, the Soviet Union, as well as the two Western powers, refrain from testing nuclear weapons."[181] (As late as October 25th, just a few days before the scheduled meetings between the three atomic powers, the Soviet Union had not confirmed its attendance. But Eisenhower sent a twenty-five person delegation to Geneva, Switzerland; the Russians showed up.)

On October 30, 1958, the White House announced "the United States will stop atomic tests tomorrow and will not resume them for one year, provided the Soviet Union carries out no nuclear tests in that period."[182] On December 18, 1958, the talks adjourned briefly "leaving most of [the] fundamental problems unresolved. . . . The most important of these concern voting procedures on the proposed control commission and the composition and functions of international inspection teams."[183] The *New York Times* sadly editorialized about the failure at Geneva. "Another effort to make the world a safer place to live—to reduce the danger, and temptation, of Atomic Pearl Harbors—has come to naught."[184]

As a consequence of the President's year-long moratorium, the Los Alamos Scientific Laboratory scientists at the Nevada Test Site wanted to complete the Hardtack series with more atomic shots and with as few delays as possible. The Atomic Energy Commission's original schedule of ten shots was increased to twenty. A *New York Times* article with the headline, "Last Nevada A-Blast Balked By Deadline," summarized the situation:

> The U.S. failed tonight, October 30, 1958, in an eleventh hour attempt to complete its current nuclear test series with a giant balloon shot. Unfavorable weather conditions delayed firing of the twentieth and last of the current nuclear test series. . . . The cancellation came only twenty minutes before the United States deadline for suspending nuclear tests.[185]

Some of the bombs were dropped from balloons during the Hardtack test series, and the press announced plans to "fire four A-tests in two days,"[186] and "to fire 5 Atom blasts."[187] The flurry of tests in October, 1958 led to radiation scares in Salt Lake City and in the downwind area, especially in the Washington County area of St. George, which had "the highest reported level of fallout in the nation."[188]

The papers ran stories about radiation and the adverse affects of "Atomic Anxieties on the Mental Health of the World,"[189] and printed letters to the editor, such as the one written by Mrs. Delome Billings on October 1, 1958, in which she urged a ban on nuclear tests and labeled the Hardtack tests "a colossal folly."[190] Maintaining that there was "no time to lose," she pleaded with readers to "send letters to congressmen and other leaders" imploring them to stop the atom bomb testing.[191] The test series ended quietly; there were no editorials in the *Deseret News*, nor did Nora Lyman's *Observations* column take note of the silence from the desert and hills to her west.

4 The Downwinders and the Trauma of Cancer, 1960–1980

The suffering doesn't seem to be going away. It keeps on going on; it's just been going on and on.   Irma Thomas, St. George, Utah

I am sad to say that my belief has been shattered by the events of recent months.   Patricia Walter, St. George, Utah

We are angry because we were used and abused by the government.
                    Janet Gordon, Cedar City, Utah

In 1960, John F. Kennedy was elected President of the United States. At age forty-three he was the youngest person ever elected to that office and his youth and experience seemed to augur well for the new decade; some in the administration likened the era to the time of Camelot, a "more congenial spot." People were being asked, by Kennedy and others, to consider what they could do to help their country. A new era seemed to be dawning in America, a time that emphasized citizen participation, political and social freedom, and an end to the twin domestic evils of the 1950s decade: McCarthyism and racism.

By 1960, McCarthyism's virulence had ended. Published in 1960, Harper Lee's Pulitzer Prize-winning novel, *To Kill a Mockingbird,* reflected a new beginning for favorable race relations in America. In 1954, segregation was legally banned by the Supreme Court and finally placed on the national political agenda. A newly aroused black population sat-in, sang-in, prayed-in, and generally stirred the conscience of the politicians and the American populace. Organizations such as CORE, NAACP, and SCLC actively sought an end to segregation and found a receptive audience in men and women in both the Kennedy White House and the Justice Department. These were practical politicans who were willing to use the powers of the federal government to assist blacks in the eradication of institutionalized racism.[1]

In 1961, Kennedy introduced Americans to the Peace Corps and to the Alliance for Progress. A goal-oriented U.S. Supreme Court, led by U.S. Chief Justice Earl Warren and Justices Black and Douglas had begun what has been referred to as a "judicial revolution" in civil liberties, civil rights and criminal justice.[2]

But the first heady days of the Kennedy presidency were marred by darker events: the embarrassing and unsuccessful "Bay of Pigs," an effort by expatriate Cubans, supported by Kennedy, to invade Cuba and overthrow Fidel Castro; the 1961 Berlin crisis, during which the Berlin Wall was erected; the Soviet Union's resumption of atmospheric nuclear testing in 1961 (followed, in September, 1961, by the resumption of atmospheric atomic testing in the United States); the 1962 nuclear confrontation between the Soviet Union and America over the Soviet's missile installation in Cuba; America's beginning involvement in a shooting war in Vietnam; the rise of a militantly nationalist and activist black group, the Black Muslims, led by the dynamic and controversial Malcolm X; and the stirring of bloody violence that was soon directed toward blacks and white civil rights workers by southern whites in Selma, Alabama and Philadelphia, Mississippi.[3]

If Pete Seeger's ballad, "Where Have All the Flowers Gone?" captured the sadness of the time, then Bob Dylan's "Blowin' in the Wind" suggested an ominous, dangerous future. Americans in 1961 and 1962 had begun to read about chemicals causing birth defects (thalidomide) and about the unintended effects of chemical pesticides (in Rachel Carson's *Silent Spring*). There was, however, an encouraging development: in August, 1963, the United States, the Soviet Union, and Great Britain signed a limited test-ban treaty that prohibited nuclear testing in the atmosphere, in space and underwater (but allowed for underground testing). At the time of the signing, over 100 atmospheric nuclear detonations had occurred at the Nevada Test Site since 1951 (Figure 12).

The deadly atomic sunburst over Hiroshima, in 1945, produced 13 kilotons of murderous heat and radioactive fallout. At least 27 of the 96 aboveground bombs detonated between 1951 and 1958 at the Nevada Test Site produced a total of over 620 kilotons of radioactive debris that fell on the downwinders (Figure 13). The radioactive isotopes mixed with the scooped-up rocks and earth of the southwestern desert lands and "lay down a swath of radioactive fallout" over Utah, Arizona, and Nevada.[4] In light of the fact that scientific research has now confirmed that any radiation exposure is dangerous, the "virtual uninhabitants" (over 100,000 people) residing in the small towns east and south of the test site were placed in potential medical jeopardy by the AEC atomic test program.

Recently, the Utah State Cancer Registry noted that although "radiation risk and atomic bomb fallout had become a public issue in Utah," it still had the lowest age-adjusted cancer mortality rate in the entire United States. Cancer incidence in Utah between 1967 and 1977 was 17% lower in males and 15% lower in females than in the rest of the United States.[5] However,

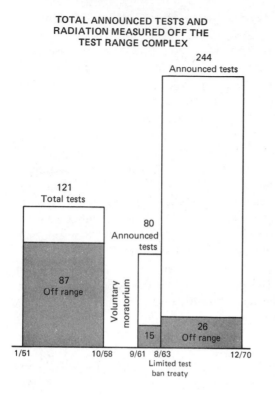

Figure 12. A chart, prepared by one of the Field Managers at the Nevada Test Site, summarizing the total number of shots, including the nuclear devices that produced radioactive fallout downwind, or "off range." (Source: Mahlon Gates, Testimony before the Commerce Committee of the House of Representatives, April 23, 1979.)

at the same time that John F. Kennedy took his oath of office in January, 1961, medical research reports began indicating an alarming increase in the number of leukemias in the small downwind communities in Nevada and Utah.

By 1961, these southern Utahns and Nevadans experienced something tragically new: clusters of acute leukemia deaths in a region of the country where deaths from these cancers were far below the national average. (Acute or sudden onset leukemia, a killer of children, develops much faster than the hard-tumor cancers. Leukemia develops in just a few years after radiation exposure, while other cancers often do not appear until decades after the exposure.)

A review of the first two Utah State Cancer Registries, published in 1972 and 1975, indicates that the acute leukemia and cancer rates in the counties closest to the Nevada Test Site increased significantly. Report Number 2, published in November, 1975, indicated that while the cancer mortality figures for the United States were higher than the corresponding mortality figures for Utah, the incidence of acute leukemia in Utah was actually higher than the incidence of acute leukemia in the United States[6] (see Table 1).

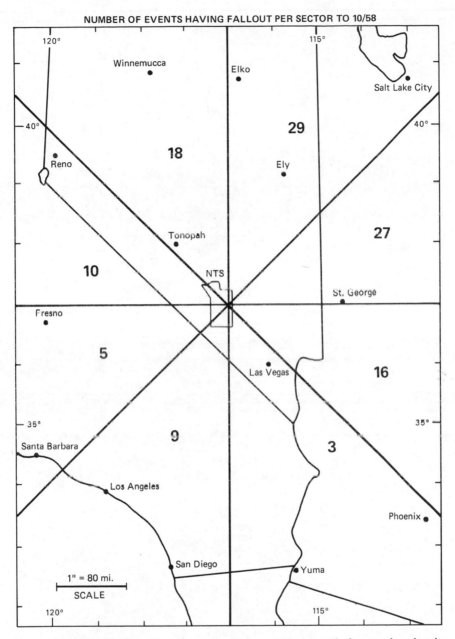

Figure 13. Plot of the number of atomic explosions ("events") that produced radio-active fallout and where the fallout came down, presented at a 1979 congressional commmittee hearing. (Source: Mahlon Gates, Testimony before the Commerce Commmittee of the U.S. House of Representatives, April 23, 1979.)

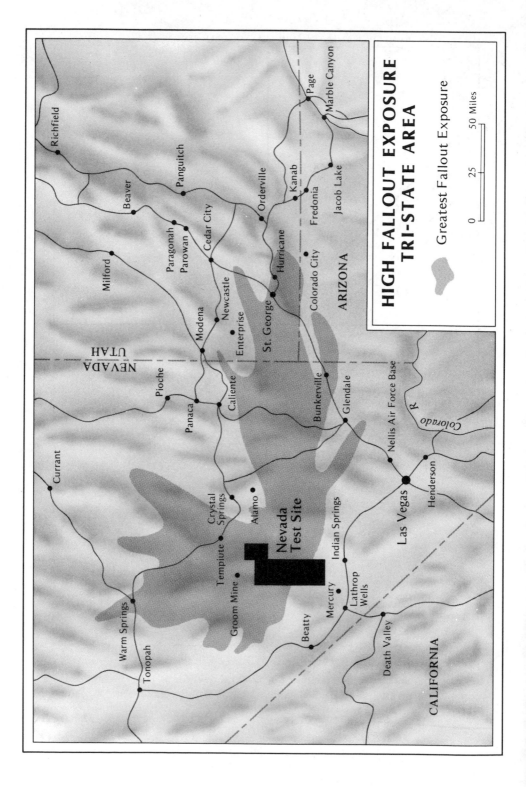

HIGH FALLOUT EXPOSURE TRI-STATE AREA

Greatest Fallout Exposure

0   25   50 Miles

Nevada Test Site

NEVADA
UTAH

ARIZONA

CALIFORNIA

Colorado R.

Richfield
Page
Marble Canyon
Panguitch
Beaver
Orderville
Kanab
Paragonah
Parowan
Cedar City
Fredonia
Jacob Lake
Milford
Hurricane
Modena
Newcastle
Enterprise
St. George
Colorado City
Pioche
Caliente
Bunkerville
Glendale
Panaca
Nellis Air Force Base
Currant
Crystal Springs
Alamo
Indian Springs
Las Vegas
Henderson
Tempiute
Groom Mine
Mercury
Lathrop Wells
Warm Springs
Beatty
Death Valley
Tonopah

Table 1. Acute leukemia and cancer rates in Utah and the United
States per 100,000 population, 1957–1974

	United States		Utah	
	Male	Female	Male	Female
Cancer mortality	174	130	133	102
Incidence of acute leukemia	4.3	2.8	4.5	3.8

Dr. Edward Weiss, of the U.S. Public Health Service, prepared a report in
1965 for the Atomic Energy Commission, entitled "Leukemia Mortality in
Southwestern Utah, 1950–1964." The report provided additional data and
an analytical judgment about leukemia deaths in Washington and Iron
Counties, the two counties closest to the Nevada Test Site. As there was no
functioning Utah State Cancer Registry until 1965, Weiss based his analysis
and judgment on data obtained from death certificates from the two counties. He noted a cluster of six leukemia deaths in 1959–1960, and observing
that the "probability of observing six or more cases in one year is less than
2%," he concluded that the "aggregation of seven cases with onset leukemia
in 1959–1960 constitutes the cluster of primary interest"[7] (see Tables 2 and
3). The Atomic Energy Commission chose not to release the Weiss report
until 1979, when its general files were opened by congressional investigations, Presidential, and Freedom of Information Act requests.

In that same year, 1965, the Utah State Division of Health published a
statistical picture of "Utah Leukemia Deaths, Per 100,000, 1956–1964."[8]
*The four Utah counties on the Nevada border—the four Utah counties closest to
the Nevada Test Site—had the highest number of leukemia deaths per 100,000.*
Iron County, Utah was first on the list with a statistic of 9.56 leukemia deaths
per 100,000 population. Millard County had 8.68; Beaver County had a leukemia death rate of 8.18; and Washington County had 8.00 leukemia deaths
per 100,000.[9]

The Downwinders' Recollections of the Testing
and the Cancers

Even without knowledge of the medical reports and Cancer Registry statistics, the residents in the small villages downwind of the Nevada Test Site
began to see and experience the pain of what they considered to be an inex-

Table 2. Observed and expected number of leukemia deaths in Washington and Iron Counties, Utah, 1950–1964

	Population		Number of Deaths	
Age	Male	Female	Expected[a]	Observed
<5	1,425	1,350	2.30	2
5–14	2,565	2,390	2.35	5
15–24	1,700	1,625	1.06	2
25–34	1,180	1,225	.83	0
35–44	1,120	1,130	1.10	1
45–54	950	880	1.70	4
55–64	750	730	2.84	6
65–74	550	475	3.99	6
75–84	230	215	2.52	2
85 and over	40	60	.52	0
TOTAL	10,510	10,080	19.21	28

[a]1950 Rates from Vital Statistics, Special Reports, vol. 49, no. 24(1959), Public Health Service NOVS.
1960 Rates from Death Rates for Malignant Neoplasms, Public Health Service publication no. 1113, December 1963.

plicable epidemic of leukemia and other cancer deaths involving their families, loved ones, and members of their extended town family.

The downwinders noted the sudden frequency—particularly among the young—of leukemias and other cancers which had been virtually unknown. Mrs. Frankie Bentley lived in Parowan, Utah, one of the many villages east of the Nevada Test Site. She recalls 1960 vividly:

> In 1960, we had four young teenagers die with leukemia, first ever in the Parowan area. . . . We have just a personal suffering of a small community for we are all involved with each other's lives. . . . The damage is done with us and it will be for a long time.[10]

Elmer Pickett referred to his own family history since the 1950s' testing:

> In my own family we have nine cancer victims, beginning with my wife who died of the disease and leukemia combined; my niece, five years old, from leukemia and from cancer; a sister, a sister-in-law, a mother-in-law, an uncle, a grandmother, and two great uncles. . . . I cannot find anywhere in our family

Table 3. Observed and expected leukemia deaths by age and type of disease in Washington and Iron Counties, Utah, 1950–1964

Age	Gran.		Lymph.		Other	
	Observed	Expected[a]	Observed	Expected[a]	Observed	Expected[a]
			Acute			
0–19	3	2.51	5	1.52	1	1.18
20–39	1	0.71	—	.12	—	.18
40–59	2	1.10	1	.20	—	.36
60 and over	—	1.02	—	.29	—	.53
TOTAL	6	5.34	6	2.13	1	2.25
			Chronic			
0–19	—	.23	—	.02	—	.38
20–39	—	.51	—	.15	—	.12
40–59	1	1.17	2	1.00	—	.52
60 and over	3	1.38	6	3.51	3[b]	1.23
TOTAL	4	3.29	8	4.68	3	2.25

[a]Based on Brooklyn, New York rates for whites from 1943 to 1952 (MacMahon).
[b]Cases 13, 19 and 25 are all questionable leukemias (see text).

records, anywhere back as far as we can go, any cancer-related deaths of any nature anywhere in our family lines. We have been a very healthy family down here. The majority of our family on both sides back through have lived to very ripe old ages . . . I cannot find anywhere in our family records of any cancer. So to me, the conclusion has to be that there was some great cause right here in this locality to bring this about in this period of time. It has all happened since the fallout. To me, the fallout is the only thing that had happened in this area that could cause such a disaster.[11]

A young student teacher in her community of Orderville, Utah, in the 1960s, and later a member of U.S. Senator Orrin Hatch's staff, remembered the large number of families "hit with medical problems." "Almost every family had had at least one member die from cancer, leukemia, or some other debilitating disease."[12] Darrell Nisson, from Washington, Utah, lost a small child to leukemia and recalled that "up to the time of this radiation from the testing, I only knew of one leukemia death in Washington of only about 450 people. . . . But since that time, there have been six cases of leukemia, including my boy who used to help me in the canyons."[13] Glenda Orton of Parowan, Utah claimed that of the 2,000 living in Parowan,

I can safely say that 30 to 40 percent of our livestockmen and outdoor people here suffered some kind of cancer since the testing. Eighteen people in my father's street have had cancer and four of these were leukemia victims. . . . I don't think there is hardly a family in Parowan who has not been touched with cancer. Our friends and neighbors everywhere have suffered. People have been very supportive, one of another.[14]

The downwind residents also recollected "the big pink cloud that hung over us for more than a day,"[15] "the times it seemed hazy in St. George due to the fallout,"[16] and how they

ate [the fallout], walked in it, breathed it, washed our clothes in it, hung our clothes out in it—very few people had driers in those times and even the little children ate the snow. You know how little kids love snow. They went out and would eat the snow. They didn't know it was going to kill them later on.[17]

Jeannie Snow, a young mother from St. George, Utah poignantly commented how she "personally remembered the pink dust in the air at the time and I huddled with all of my little ones around me. But what could you do when you can't see [the fallout], you can't feel, you can't taste, you can't hear?", she asked recently.[18]

Gloria Gregerson, who lived in Bunkerville, Nevada, recalled the nightmarish developments in her small town: "Of 50 families I know there, only four have no cancer in the family. One family's members have endured seven cancers and 12 miscarriages."[19] Gregerson recollected an incident that occurred at the beginning of the atomic testing program when she was twelve years old:

I remember playing under the oleander trees, which is a wide-leaf tree, and the fallout was so thick it was like snow . . . [We] liked to play under the trees and shake this fallout onto our heads and our bodies, thinking that we were playing in the snow. I remember writing my name on the car because the fallout dust was so thick. Then I would go home and eat. If my mother caught me as a young child, I would wash my hands; if not, then I would eat with the fallout on my hands.[20]

Four years later, Gregerson, then seventeen, was diagnosed as having ovarian cancer, then cancer of the intestines and stomach, resulting in thirteen operations. In 1970, she developed skin cancer in the vaginal area, and in 1979, leukemia. She was in remission in 1981, but needed a bone marrow transplant. She died in March, 1983, at the age of 42.

The comment of Mrs. Irene McEwen of Panguitch, Utah captured the prevailing perception of the small-town residents in Utah, Arizona, and Nevada who have testified before congressional committees and stood as witnesses in federal court negligence trials: "there isn't a block where it hasn't hit . . . House after house, street after street, over and over again. It strikes—and keeps on striking.""[21]

These recollections are tinged with bitterness toward the federal government, in whom they previously trusted with patriotic loyalty. There is sad irony in the fact that these loyal citizens were, in Nisson's words, "expendable." He added, "We got in that mess over in Korea and were afraid we were going to get into an atomic war over there, so there was a crash program and I guess the government felt it was necessary. But they sure didn't have a regard for us citizens around this area."[22] Glenda Orton echoed his views: "We trusted the government when they told us that it was safe and so we didn't take precautions that we might have otherwise done had they told us the methods to protect ourselves."[23] Mrs. Bentley was blunt: "Our government lied to us, they didn't think that we were intelligent enough to have taken the precautions."[24]

Coping with the Cancers

Coping is the human effort to deal with stress. Usually a person employs one of two basic coping mechanisms in response to stress or trauma: (1) "Address the source of the stress or reduce or remove the threat or danger that is posed," or (2) "focus inward and attempt to control the fear and emotional responses associated with exposure to stress."[25] The small-town residents who lived downwind of the Nevada Test Site were helpless to remove the threat of cancer since statistical cancer and leukemia clusters had already developed.[26] Instead, they had to react to the perceived adverse consequences of the trauma.

In such a traumatic situation, a person may choose to do nothing: Helplessness is one possible response to the crises that confronted the downwinders in these communities. Social scientists studying people's response to stress have noted that a "loss of perceived control over circumstances often leads to a . . . feeling of helplessness."[27] Certainly, when a person confronts the trauma of cancer in the family or larger community, "views of living in an orderly, understandable, meaningful world" are often shattered.[28] Another response is to seek information, or to associate with others to battle the consequences of the stress: "one can choose to alter the setting."[29] The person who attempts to alter the setting is "influenced by the . . . ability to find meaning or purpose in his or her misfortune." Indeed, some of the downwinders "seemed compelled to make sense of their (unfortunate) experiences,"[30] and sought to find meaning in the crisis.[31]

Social psychologists who have researched the chronic stress generated by the technological, near-disaster at Three-Mile Island in 1978 have concluded that social supports available to the Three-Mile Islanders "bolstered" the residents' ability to cope with the crisis and minimized the aversiveness of stress. But, these researchers found that these supports did not "allow either termination of stress or ready adaption to it."[32] Similarly, the religious, patriotic citizens in these small towns downwind of the Nevada Test Site needed moral support in their struggle with the resultant cancers and leukemias, and found that their communal ties to the Mormon church "facilitated coping . . . and reduced the psychological and behavioral consequences of stress."[33] But, the group support could not eradicate the anguish and suffering that accompanied the chronic stress.

For the Utah, Nevada, and Arizona residents of the downwind communities, emotional survival depended upon their accepting the reasons for the malignancies. Victor Frankl has written that the most powerful human motivation is the search for meaning in injustice and senselessness.[34] As the cancer deaths continued, the downwinders very gradually developed in their hearts and minds a conviction that the Atomic Energy Commission was responsible for the deaths in their communities and for their suffering in the wake of the loss of their loved ones.

Trauma of this kind is often lifelong. For example, a sizable minority of persons who experience trauma, including the trauma of loved ones dying of cancer, never accept or recover from their distress.[35] "Our loss is, was, and will always be an emptiness in our hearts and immeasurable in heartache and anguish that all the years cannot erase," said Mrs. Dorothy Fox[36] of Vegas Heights, Nevada, who lost her five-year-old son to leukemia. She will always believe that the death occurred because her son was continually exposed to the AEC-produced radioactive deposits that fell to earth after the atomic bursts.

The downwinders' distress "stems not only from [the terrible act] but also from the resultant feeling that people cannot be trusted."[37] The collective feeling of distrust felt by this large cohort of religious, conservative and patriotic Americans has altered their view of the world, especially government and politics. "We believed them [AEC officials] . . . because whatever they said, we thought was true," said a resident of St. George, Utah.[38] The downwinders perceived that the Atomic Energy Commission had committed a breach of faith;[39] "they lied to us," exclaimed Alta Primm of St. George, whose experience has fostered in her a deep and unrelenting bitterness toward government in general. To hear citizens who "are as patriotic, industrious, religious, conservative and honest a people as can be found anywhere in this country"[40] call the governmental leaders liars is to lend credence to the view that "one of the most devastating aspects" of the downwinders' continuing traumatic experience is the fundamental alteration of their world view.[41]

It is difficult for the researcher to put together an overarching statistical portrait of the malignancies that developed in the downwind communities during the past few decades, because a central clearing house for these statistics was unavailable until a decade or so later. The Utah Cancer Registry did not begin systematically compiling cancer mortality information until the late 1960s.

The only extant records are the death certificates on file in the local town halls of these small communities. There are problems associated with the use of these medical records, including misdiagnoses and certificates that were never filed. Medical epidemiologists who have written about cancer mortality in southern Utah, as will be shown in the next chapter, have had to rely almost exclusively on these scattered death certificates and on the recollections of the downwinders themselves.

However, most medical epidemiologists researching the basic medical/ethical riddle of the increased numbers of cancers in the communities downwind of the Nevada Test Site have determined that an excess of cancers and leukemias did occur. For example, the most recent study, published in 1984 in the *Journal of the American Medical Association*, which based its findings on a 1981 study of over 4,000 Mormons who lived in the downwind communities in 1951, found an *excess* of 109 cases of leukemia (179 statistically expected, 288 actually observed) in the downwinder communities![42]

One of the ways in which the downwind citizens in these small towns responded was by forming and supporting pressure groups in an effort to achieve justice for the victims of this alleged government-perpetrated, medical calamity. In addition, citizens attempted to stop all nuclear testing in order to prevent these traumatic events from recurring. In trying to make sense of their experiences with the malignancies, a number of the downwinders took action they would not have contemplated before their personal and communal tragedies. "Many people come to view their aversive experience from a purposeful or meaningful perspective."[43] Three people who were moved to action after they concluded that their government had acted in a negligent manner are Janet Gordon, Preston Truman and Bennie Levy.

Janet Gordon and Citizens Call

Janet Gordon, a native of Southern Utah, organized Citizens Call in 1978. She and Elisabeth Bruhn Wright, along with Preston Truman and Mary Lou Milburg, were among the small group of downwind victims and concerned family members "white-hot with anger," as Janet recently stated.

In testimony before a congressional committee in 1981, Janet recalled that her brother Kent

> was exposed to fallout in the early 1950s while working on the range [in southern Utah] with my father. . . . When Kent's horse died a few weeks later after some particularly dirty test, we made no connection between radiation and his death. When Ken died a painful death from pancreatic cancer at the age of 27 in 1963, we didn't attribute it to his radiation exposure, because we didn't know what the hazards of radiation were. But now, we do know—my family and I firmly believe that Kent died because of the government's failure to warn us about the dangers. And my family is not alone."[44]

She stated that the people downwind are bitter and angry, "because the government did not tell us what could happen to us in the 1950s." Janet concluded her testimony by stating, again, that "we are angry" because the government was playing God. "Does the government have the right to play God? Who will die and who will live?"[45]

It was after the 1978 congressional hearings on the Nevada Test Site action in the 1950s that she and others concluded that the federal government had been telling them lies. Janet and Elisabeth decided that their small group should begin to "call citizens" to inform them about the reasons for their malignancies, and to provide them with the necessary cancer screening and transportation to these screening centers (often many miles away from the downwinders living in isolated rural areas).

The group decided to use this collectivist Mormon social concept as the name of their organization: Citizens Call was born to provide assistance to the people of southern Utah, Nevada, and Arizona in their hard struggles with the cancers (see Appendix C). From its beginning in 1979, the organization's basic goals were clearly defined ones: (1) to provide a mobile health screening program for the downwinders; (2) to provide health care for victims of cancer and leukemia in southern Utah and Nevada; (3) to organize a grassroots information dissemination network among the 100,000 persons who lived downwind of the Nevada Test Site; (4) to gain additional information from the government and independent analysis of the data turned over by the Nuclear Regulatory Commission, Veterans Administration, and other federal agencies associated with the nuclear testing program; (5) to actively campaign for a comprehensive test ban treaty between the United States and the Soviet Union; and (6) to continuously monitor the underground testing program.

"We must educate ourselves and the public," said Janet Gordon recently. "We must seek assistance for the victims and support them in litigation. We must make the public and the politicians understand and we must provide tangible support (hospice) for the victims."[46] Supported by funds raised by Citizens Call, the Hospice of Southwestern Utah (HSWU) was recently opened in Iron County, Utah to assist the terminally ill and their families. In 1982, Paul Carpino joined Citizens Call to develop the grassroots network that linked downwinders with medical and legal help, but he has recently left the group because of its endemic problem: lack of funds. Since Carpino's departure, Gloria Gregerson's recent death, and Jay Truman's move to the another organization, the *Downwinders*, Janet Gordon is the only staff person working for Citizens Call in 1984. "I need to raise money for this work," she said recently, "even if it means my waiting on tables again."[47]

The group has also called for

> assessment of such actual data as exists, including biological indicators, conducted by independent scientists without a vested interest. We believe that everyone associated with and about a test site does have a vested interest.[48]

In addition, since the association of radiation exposure and problems such as birth defects has been documented,

> Justice absolutely dictates that . . . extensive independent epidemiological research should be done with the goal of understanding cumulative effects of low-level radiation exposure.[49]

Citizens Call is committed to a comprehensive test ban treaty that would end all atomic testing. Through their publication, *Citizens Voice*, the organization has campaigned to end the testing, and it has expanded its monitoring efforts to include the health problems of uranium miners and nuclear munitions plants workers, such as those working at the Rocky Flats, Colorado plant.

In addition, Gordon spoke out for justice for the Paiute Indians when she appeared before a congressional committee recently:

> These people tended to live on sometimes poached deer and rabbits and pine nuts during those [above-ground testing] days. Their exposure had to be at least as great as any of the rest of us. A lot of them did not even have the benefit of food coming in from the outside, as some of us did when we went to the store to buy our food . . . Some fair remedy must be made.[50]

Janet Gordon and the group also believe that "all of the victims deserve redress, including the veterans."[51]

Since 1980, candlelight vigils have been held at the Nevada Test Site and other sites across the country on January 27 (the anniversary of the first above-ground shot set off at the site in 1951). The purpose of this vigil, according to Janet, is "to remember and honor victims of radiation from nuclear weapons testing programs, as well as the victims who have been exposed to nuclear fallout and other radiation along the nuclear weapons cycle," i.e., radiation effects from uranium mining and milling, weapons fabrication, transportation, and storage of nuclear wastes.[52] Since the group's organizational activities began in 1978, chapters have developed in Nevada, Washington, D.C., and California, and coalitions have been organized between Citizens Call and other groups to educate the public about the hazards of nuclear testing. A Radiation Roundtable was held in Washington, D.C. in the spring of 1983, and Citizens Call helped form Western Solidarity, another concerned citizens' group that operates out of Denver, Colorado.[53] Janet Gordon's Citizens Call group was and is one response to the tragedy confronted by the downwind citizens.

Preston Truman and the Downwinders

Another small, struggling, concerned citizens' group was formed in Salt Lake City in 1980 by one of the former directors of Citizens Call, Preston (Jay) Truman.

Jay Truman is a young man from southern Utah who vividly remembers the above-ground testing era. He can still visualize the flash "as it would light up the mountains around town, sounds of the blast as they would ripple through the hills, and of course, the clouds from the tests as they drifted by later that day."[54] Truman, who grew up in the small southwestern Utah town of Enterprise, copes with the cancer and fallout traumas by working tirelessly toward an end to all nuclear testing around the globe. His resolve was strengthened further when he attended his high school reunion in Enterprise a few years ago. Part of the reunion was held at the local cemetery. "Out of nine young friends he had grown up with in the small rural community of Enterprise, Truman was the only one to reach his twenty-eighth birthday— the rest had died of cancer or leukemia."[55] He is consumed with the desire to end all nuclear testing through the signing of a comprehensive test ban treaty.

Truman and his organization are concerned about the new generation of nuclear weapons—the MX, Cruise missile, neutron bomb—which have been developed since the tests went underground. He believes the problem is compounded by the fact that the testing is now underground and therefore "not so visible and few will notice a 30% increase in testing in 1982."[56] In the newsletter (see Appendix C) and pamphlets he produces and distributes, he implores, "Write the final chapter on the proliferation of nuclear armaments, because," he says, "with thousands of nuclear weapons deployed around the world today, we all live downwind."[57]

Two of the group's booklets illustrate the central purpose of the Downwinders. *A Primer on Nuclear Testing: Everything You've Always Wanted to Know But Have Never Been Told*, written by Truman and Dennis Bedolla, a concerned activist from California, presents a fairly comprehensive and easy-to-read elaboration of America's nuclear testing program from its beginning at Alamogordo, New Mexico in 1945, to the present-day underground testing of nuclear weapons. The booklet also provides a good discussion of what Truman refers to as the "Testing Complex": the scientific laboratories—Lawrence Livermore (Berkeley, California); Los Alamos (New Mexico); and the Sandia Laboratory (Albuquerque, New Mexico); the industrial plants, such as Rocky Flats, Colorado, where the parts of the weapons systems are made; and the Nevada Test Site, where all the parts, as well as the scientists and the military persons came together for the events—the shots.

A second small booklet, entitled *A Guide to a C.T.B., Comprehensive Test Ban*, educates the reader on the dynamics of the nuclear testing programs in the United States and the Soviet Union and explains the vital need for a comprehensive test ban. Simply stated, Truman's message is that such a test ban "would end the long nightmare of radioactive fallout that has followed nuclear testing since the first mushroom cloud rose above the New Mexico desert in July, 1945."[58]

Finally, Truman's organization has joined with other groups (who share the commitment to end nuclear testing) in providing petitions for public signatures which are then forwarded to officials in Washington, D.C. As the leader of this small group of Downwinders, Truman has acted in a way that is consistent with the ethical imperative inherent in the Nuremburg trials of Nazi leaders at the conclusion of World War II. The trials, he said, "made it clear that one has a moral obligation to disobey orders" and to fight to end governmental actions that lead to ethically bankrupt conclusions and needless deaths of innocent civilians.[59]

Bennie Levy and the Nevada Test Site Workers Victims' Association

Bennie Levy, a veteran of World War II and an iron worker at the Nevada Test Site from 1951 to 1977, is one of over 200 former Nevada Test Site workers who still live with cancer and leukemia; over 100 others have died of these maladies. In 1980, Levy organized the Nevada Test Site Workers Victims' Association in an aggressive effort to seek compensation for the workers' widows. "I'm fighting my own country," he bitterly stated recently, "because they're wronging my people. The government did not give them the protection at the NTS."[60]

As an iron worker at the Nevada Test Site, Bennie went to ground zero less than an hour after the detonation "while the ground was still smoldering,"[61] to recover the instruments for the scientists who worked at Livermore and Los Alamos laboratories. He trusted the government officials who told him and his co-workers "it's safe to go in," even though he was not wearing any special protective clothing.

Today, Levy is angry about the government's duplicity and believes that the government knows more about the actual radiation doses than it has revealed to the public in recent years. "They used these people as guinea pigs and they know how much they were exposed to and what has happened to them since," he said recently.[62] He also charged that the Atomic Energy Commission and other government agencies "knew exactly how much actual exposure was received by the test site people, but . . . they altered the information from 1963 to 1965."[63]

Levy's skepticism developed during the early days of the above-ground testing program. Testifying recently before a congressional committee, he recalled that one of the Nevada Test Site workers, Oral Epply, had gone to ground zero minutes after a bomb blast and died a few days later. But no on-site investigation was ever performed by the AEC, Levy testified. Furthermore, after Epply's organs were removed and transported for autopsy to the AEC laboratory at Los Alamos, New Mexico the doctors there denied ever receiving them. Levy's review of the Epply hospital records and autopsy report confirmed the fact that the organs were sent to Los Alamos, and yet the autopsy report has never been uncovered through use of Freedom of Information Act.

Levy is also angry at the Atomic Energy Commission and Department of Energy for their negligence and indifference to the problems encountered by the Nevada Test Site workers and downwind persons. At the 1982 hearings, Levy recalled a situation involving John Numerousky, who was 37 when he came to work for the Atomic Energy Commission at the site in 1957. He worked at the Nevada Test Site for ten years in the underground tunnels, and came into close contact with the radioactive earth. After coming down with skin cancer, he left the Nevada Test Site. Nine years later he developed leukemia, and he died four years afterward. Numerousky's widow was destitute and had to leave the area for Oklahoma after the Department of Energy refused to consider whether the government was responsible for her husband's death. Bennie Levy's convictions and bitter memories have propelled him to seek help, through his organization, for the widows and children of the deceased Nevada Test Site workers.[64]

The goal of his organization is to provide information about the actual Nevada Test Site testing situation to Washington, D.C. legislators and to federal judges who hear tort liability cases about Nevada Test Site testing and the resultant illnesses and deaths from low-level radiation fallout. In working with fellow Nevada Test Site workers who have cancer and who also happen to live downwind, Levy is pressing for additional public information about the actual doses received by the "guinea pigs" and is trying to assist the widows of the deceased Nevada Test Site workers.

Why Is It Happening to Us?

There is little doubt that radiation can cause cancer. However, for the person who contracts the frightening and often deadly disease, "the link between a given radioactive exposure [and the tumor] can only be drawn statistically."[65] For medical researchers and medical epidemiologists, as well as for lawyers who argue their cases in a federal court, this is a difficult statistical problem to resolve. As will be shown in subsequent chapters, it is a problem that is being wrestled with by medical researchers, medical demographers, lawyers, federal judges, federal administrators and national legislators.

However, the persons living downwind in these predominantly Mormon villages felt certain that the cancer and leukemia epidemics occurring in their small towns since the late 1950s were linked to the negligent actions of the AEC and its successor federal agencies who developed and implemented the continental nuclear testing program. The question "Why is it happening to us?", is answered by pointing to the careless actions of the Nevada Test Site operating personnel. Janet Gordon, a resident of Cedar City and the director of Citizens Call, bluntly stated before a 1982 congressional committee: "We became and we remain angry [because we] feel that we were used and abused" by the federal government.[66]

5 Associations Between Radiation Exposure from Nuclear Fallout and Cancer: The Medical Controversy, 1961–1985

When experts disagree, whom do we believe?
John Gofman, M.D.

Statistical models in my view should mirror reality as far as is feasible.
L.D. Hamilton, M.D.

We are the statistics. They are not looking at us. When are they going to look at us?
Janet Gordon, Cedar City, Utah

The Probable Pathogenesis of Radiation-Induced Cancers

There are billions of cells in the human body, each a few thousandths of a millimeter in diameter, which are filled with molecules containing acids, proteins, fats, and carbohydrates. The nucleus (Figure 14), the core of each human cell, contains our genetic information, stored in the deoxyribonucleic acid (DNA) molecule. "The discovery of DNA, by Bernard Crick and James D. Watson, with all its biological implications, has been one of the major scientific events of this century," commented Sir Lawrence Bragg in 1969.[1] "The DNA represents the information base for the enormous biochemical capabilities of the cell, including instructions for carrying out the remarkable process of mitosis," the formation of new cells in the body.[2] In his book, *Double Helix*, Watson recollected his thoughts about this discovery that illuminated the physiochemical nature of this gene: "The existence of two intertwined chains with identical base sequences . . . strongly suggests that one chain in each molecule had at some earlier stage served as the template for the synthesis of the other chain. Under this scheme, gene replication starts with the separation of its two identical chains. Then, two new daughter strands are made on the two parental templates, thereby forming two DNA molecules identical to the original molecule" (Figure 15).[3]

Ionizing radiation is a carcinogenic agent that has been proven to cause cancer and leukemia in humans. Radiation can also cause genetic damage when its swift-moving energy bombards our cells and tissues. Epidemiological studies clearly show that a population exposed to a certain dose of radiation, however low, will show a greater incidence of cancer than an unexposed population of the same size.

With the DNA molecule at their epicenters, normal cells function as team players, according to John Gofman.[4] In humans, cells divide only when the body's integrity, growth and health necessitate it. Under controlled circumstances, i.e. "cellular control,"[5] mitosis (cell division) produces two daughter cells with an identical complement of 46 chromosomes. All these cells will possess copies of the DNA molecules of the parent cell. This process occurs daily in our bodies as part of the DNA's cellular control function.

Injury to our cells can cause "preconception injuries," i.e., genetic damage and mutations such as cleft palate, Downs Syndrome, and other genetic abnormalities in unborn offspring. Furthermore, cell damage can result in chronic ailments such as premature aging, nerve cell damage, diseases of old age, paralysis, and loss of faculties.[6] However, by far the most immediate and possibly the most notorious result of cell and tissue injury is cancer.

While medical scientists do not know the precise series of events that takes place in the development of cancer, or "the key intracellular event that sets a certain organ or tissue on the path to cancer development,"[7] there is

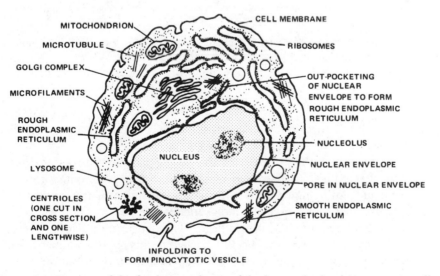

Figure 14. Diagram of the fundamental unit of the human body, the eucaryotic cell. (Source: GAO, *Problems in Assessing The Cancer Risks in Low Level Ionizing Radiation,* pp. 2–19.)

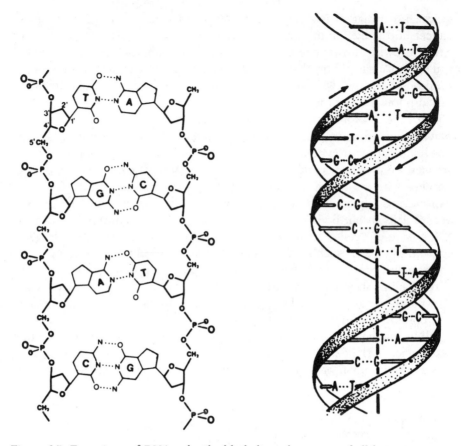

Figure 15. Two views of DNA—the double helix—the source of all human genetic characteristics and the structure that can be adversely affected by ionizing radiation. (Source: GAO, *Problems in Assessing the Cancer Risks in Low Level Ionizing Radiation,* pp. 2–21.)

general consensus that human cancers most likely develop from events that initially take place in a single cell of the human body.

One way in which the normal growth and regeneration processes can be adversely affected is if an energized radiation ray or particle enters the biological picture. All living tissue is sensitive to ionizing radiation and can be damaged or killed by it. Especially radiosensitive are embryos, fetuses, young children, and women of child-bearing age.[8] The danger of radiation, even at low levels, is that the tissues and cells exposed to this energy bombardment, if they are not killed outright, may undergo radical transformation due to cellular damage.

Radiated rays can rip electrons from a cell's atoms, or violently disrupt its molecules, upsetting the delicate balance within our cells and tissues. Once ionized, the atoms freed by the radiation can create havoc in the DNA molecule. The result can be one of three possibilities: (1) inactivation of the cell (its death); (2) cell mutation—inheritable alterations; or (3) transformation—visible alteration in the chromosomal makeup of the DNA (Figure 16). The ionized atom "shears off" part of a chromosome, causing serious biological damage in the region of the cut. For example, chronic myelogenous leukemia is characterized by cells that demonstrate an apparent deletion or cut of a part of a long arm.[9]

While the body does have defenses, i.e., internal repair mechanisms, normal cell suppression of the growth of malignant cells, and an immune system (bands of leukocytes that seek out and destroy invading, dangerous microorganisms) the leukocytes themselves are very sensitive to ionizing radiation. The invading, energized ionizing radiation ray or particles frequently kill the leukocytes and then do their damage to the cells, unimpeded by the body's immune defenses. Without the defenses, the cells are open to injury from the radiation in the form of cutting and shearing.

When the ionized atom damages the mechanism by which the cell controls its own reproduction, causing the cell to multiply too rapidly, that single cell may become the seed cell for a cancer. "Cancer is the disturbance of this control function, resulting in a proliferation of cells that are not only not needed, but ultimately cost the life of the host individual. Cancer cells keep on dividing when there is no need for them to do so."[10] This carcinogenic process is one of the ways the body responds to basic "biological insults."

In its early stages this cancerous development in the human body is undetected by either the victim or physician because the change in cellular development occurs in an area no more than 10 to 20 micrometers large.[11] A

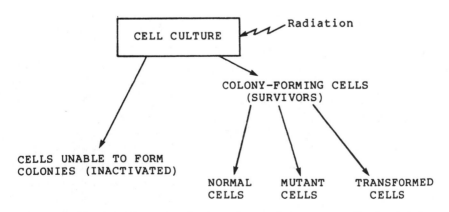

Figure 16. Schematic diagram explaining what can happen to a cell when it is bombarded by radiation. (Source: GAO, *Problems in Assessing the Cancer Risks in Low Level Ionizing Radiation,* Vol. 2, pp. 5–3.)

renegade, supportive vascularization process arises to provide the cancerous cell with blood and nourishment. Given the fact that in a single gram of tissue there are 100 million to 1 billion cells, it requires the presence of a mass of between 10 and 1,000 grams of cancerous (1 to 10 billion) cells before a medical diagnosis of cancer can be confirmed. Through the destruction of healthy cells adjacent to it, the cancerous support system is, by that time, well defined, having found a site in the human body in which to nourish itself and grow.

Because, theoretically, a single, radiated energy ray or particle can trigger this runaway alignment of cell division, most medical researchers support the assertion that exposure to low levels of radiation can harm humans. There is also agreement among cancer researchers that "there is *no* harmless dose of radiation; every dose adds to one's *risk* of cancer. And, risk means probability."[12] Consistent with these views is the belief that a person's age at the time of irradiation is important, i.e., that unborn fetuses and young children are far more radiosensitive than are mature men in their fifties. Today, the scientific dispute that rages in the medical journals; in testimony before congressional committees, workman's compensation boards, and federal judges hearing radiation injury tort cases, and at professional meetings about cancer and radiation, concerns not the correlation between the two but (1) the extent of the risk, (2) the degree of probability of cancer arising from human exposure to the radiation produced by the above-ground tests, and (3) the way in which a medical researcher ascertains risk and probability.

The General Contours of the Medical Controversy

There is no doubt that radiation can cause cancer. "For an individual, however, the link between a given radiation exposure and subsequent (5 to 50 years) development of a cancer can only be drawn statistically. That's because, to the physician, a radiation-induced cancer is indistinguishable from a cancer caused by work place chemicals, by natural toxins, or possibly even by defective genes."[13]

Furthermore, our empirical knowledge about the effects of low-level radiation exposure on humans is almost nonexistent because of science's inability to develop extended medical studies of populations affected by low levels of radiation exposure, including autopsy evaluations by medical researchers.[14] At this time, it is difficult, if not impossible, to obtain reliable cancer-incidence data for low doses of radiation. "For purposes of risk estimation, dose-response relationships observed at high doses must necessarily be extrapolated into the low dose region."[15] For some medical researchers, therefore, the question of *risk* to human health from continuous exposures to low levels of radiation is still a speculative one.

Medical researchers *estimate* the *risks* of contracting cancer and leukemia from low doses (0–10 rads) of radiation "primarily from observations at relatively high doses in human populations exposed to nuclear explosions and medical radiation."[16] They must also try, where possible, to draw associations between an adverse health effect in a population (cancer, for example) and possible carcinogenic causes (radioactive fallout from the Nevada Test Site.) Naturally, scientific disagreement has arisen concerning the scientific methods used in drawing these associations. Such an estimate of risk is fraught with methodological dangers and sure to be criticized intensely by those whose scientific studies show no such correlations.[17]

For both the individual who develops cancer and the population exposed to low levels of radiation, answering the question "why does this cancer happen to me/to my community" means relying on imprecise medical epidemiological studies that attempt, through reconstruction of time and place to measure the amounts of radiation received by downwinders and to identify the environmental and biological conditions that might have been the cause of the malady. (See Appendix B.)

Epidemiological Studies of Low Levels of Fallout and the Incidence of Cancer and Leukemia, 1961–1985

There are a number of specific problems associated with determining the amount of radiation that people living in the area downwind of the Nevada Test Site received when the atomic detonations scattered the radiation on these small towns and the grazing lands. These problems surfaced as various medical studies attempted to use government-generated information and found out that it was extremely unreliable. First of all, there are many persons—downwinders, medical researchers, politicians—who believe that the available data is unreliable.

Dr. Harold Knapp, a government medical researcher, insisted recently at a Dose Assessment Advisory Group (U.S. Department of Energy) meeting in Las Vegas, Nevada, that the downwinders "folklore" is so strong as to suggest to him that there were much higher dose readings than those produced by the Atomic Energy Commission.[18] There may very well be (as yet, undisclosed) depositions taken by Atomic Energy Commission and Public Health Service personnel from ranchers and sheep herders and others who received whole-body radiation doses far in excess of the AEC records, Knapp speculated.

Beyond intentional tampering with the spotty records released by the Atomic Energy Commission, there are questions associated with possible original underestimation of dosage statistics, inaccuracy of fallout maps prepared by the AEC, poor cloud tracking, etc. Furthermore, according to a 1980 Federal Government report, the accuracy of the monitoring instruments themselves is suspect and may have been off by as much as 30%. In a given monitoring operation, if they were in the right spot as the radiated particles drifted down to earth, the Public Health Service monitors took readings in their cars from roads and areas immediately adjacent to the highways. Yet, in a given area, there were great variations ("hot spots" and "cold spots") of fallout materials, which may not have been recorded by the monitors. There was a plus or minus factor of 2 associated with *all* off-site radiation readings.[19]

Utah cancer researchers point out that the dose data regarding the Hiroshima atomic bomb have recently been reviewed and revised because of later analyses of geography, and wind direction, and other conditions at the time of the detonation. These researchers also suggest the possibility of reviewing and re-estimating the actual doses that fell on the downwind persons.[20]

Another dilemma for medical scientists exploring the possible associations between cancer and the downwinders' exposure to low-level radiation is that through the end of the 1958 testing period, the ingestion of radioactive particles that may have appeared in the food chain was not being measured.

Furthermore, dosimetry data are gross estimates at best about radiation fallout in a specific time and place. Without more specific information regarding the whereabouts of a resident in St. George, Utah during the time of the fallout, or why a *particular* person died from cancer or leukemia, the data are of limited value.

Individual differences, such as age at time of exposure, possible shielding from the radiation, health, sensitivity to radiation, as well as environmental variability are not factored into the quantitative models and are not reflected in the dosimetry data.

The men and women who engage in this cancer research are called medical epidemiologists. They study health distributions in human populations, as well as the determinants, i.e., any measurable factor, event, or characteristic, of the population's state of health. The substance of an epidemiological study involves a careful examination of (1) the occurrence/evolution of a health disorder, (2) the etiology or causes of the health disorder and, finally (3) the control/prevention of the health disorder.

While some medical epidemiologists work in a controlled laboratory environment or do case studies, the basic approach to epidemiology is the population study, which attempts to link the disorder with the biologic and environmental habitat. Essentially, it is an analytical survey which observes a population, e.g., a population of Utahns exposed to low-level radiation in the form of fallout, and compares it with a control population who has received no low-level radiation exposure.[21] Medical epidemiologists who search for answers to the "association" question involving leukemias and radioactive fallout conduct this type of survey which "can exercise no direct control over the variables that are the object of study; it must exploit and observe the unfolding of the natural and the social environment through history."[22]

A number of important small-scale studies, relevant to the issue of cancer incidence in Utah, Nevada, and Arizona, have been performed by scientists since 1961, and many of the findings are worth noting.

1. The Weiss Study (1961): Leukemia in Utah

There have been at least nine sets of studies by medical epidemiologists concerned with the relationship between the atomic testing at the Nevada Test Site and its possible adverse impact on the health of residents in areas of the highest fallout exposure.[23] One study, by Dr. Edward Weiss, a Public Health Service medical researcher, was concluded in 1961. For a number of fundamental *policy reasons*, the AEC did not publish his report, partly because of "potentially detrimental effects upon the government's nuclear weapons testing program,"[24] according to congressional researchers. The refusal to circulate the Weiss Report was evidently due to fears that such a preliminary medical statement would trigger a negative congressional reaction and would wreak havoc with the AEC testing program. A 1965 letter from Gordon M. Dunning, Acting Director, Division of Operations, AEC, to Dwight Ink, Assistant General Manager, AEC, sums up the feelings that the AEC bureaucrats had toward Weiss and other medical researchers who suggested the possibility of associations between radioactive fallout and the increased incidence of leukemia in southwestern Utah. "There were so many fallacies in the Weiss report," wrote Dunning, "that I suggested that other reasons for performing the studies were more valid than fallout." Weiss's study contained "unwarranted conclusions." Such studies "will be accompanied by much fanfare and publicity in Utah. Probably several M.D.s from Washington will appear on the local scene in southern Utah, establish a headquarters, and use local school facilities to examine 2,000 school students. All of these will be done under the banner of fallout from the Nevada test," complained Dunning to Ink.[25] The study was finally released to the public in late 1978.

Weiss's study, "Leukemia Mortality Studies in Southwest Utah," reported for the very first time a very high increase in leukemia deaths in southern

Utah. Specifically, he confirmed 25 leukemia deaths (the expected mortality figure was 19) in two Utah counties, Washington and Iron, between 1950 and the time of the study.

2. The Knapp Study (1963): Radioactive Iodine
in the Food Chain

Dr. Harold Knapp's contribution to the medical literature was to focus, for the first time, on the introduction of radioactive iodine–131 into the food chain of the people living in high exposure areas downwind of the Nevada Test Site, an issue that had not interested AEC medical researchers. His study was entitled, "Observed Relations Between the Deposition Level of Fresh Fission Products from Nevada Tests and the Resulting Levels of I–131 in Fresh Milk."

After battling lower echelons of the medical hierarchy in the AEC to allow him to publish his results, Knapp, on September 12, 1962, wrote to Charles Dunham (director of the AEC's Branch of Biology and Medicine) that "it would be a mistake for the AEC to attempt to withold the report from other government agencies pending further study. If the conclusions are incorrect or misleading at this stage, then they can all be blamed on me; if they are essentially correct we at least have the advantage of telling unpleasant news ourselves, and we are not vulnerable to a charge of having supressed or misrepresented information on fallout, or worse still, of not being competent to find our own problems."[26] Dunham's initial response to Knapp's study was to label it "amateurish to a degree. Its inadequacies, i.e., almost total absence of thoughtful basis for many sweeping assumptions, uncritical use of analogy and models, are such as to make me reluctant to waste any more people's time on this draft."[27] The Atomic Energy Commission finally published Knapp's study, but as a *confidential* report that was not made available until 1978, when the files were open due to Freedom of Information Act requests from newspapers.

3. The Center for Disease Control Study (1965):
Utah Leukemia Clusters

In 1965, the Center for Disease Control published its findings concerning a series of investigations of leukemia clusters in the tri-state area surrounding the Nevada Test Site, as well as other parts of the country. The Center for Disease Control identified a number of small cities in the high exposure area east of the Nevada Test Site, specifically, Fredonia, Arizona, and Monticello, Parowan, and Paragonah, Utah, that had a much higher observed number of leukemias—"several times the expected rate of leukemia."[28] The CDC study, however, did not speculate on the cause of these cancer clusters in the downwind region just a few years after the above-ground atomic tests ended.

4. The Rallison Study (1974): Thyroid Cancer

The first major, nongovernmental report on cancer and fallout was conducted by a University of Utah medical researcher and his associates between 1965 and 1970, and was published in 1974.[29] Dr. Marvin Rallison and his colleagues focused on the question of increased thyroid cancers in children in the high exposure areas east of the Nevada Test Site.

Concerned about the children who had ingested milk irradiated by iodine–131, Rallison began his studies in 1965, searching for thyroidal abnormalities. Between 1965 and 1971, his group annually examined the thyroids of thousands of school children in two heavily exposed counties, Washington County, Utah, and Lincoln County, Nevada, and in one remote control population county, Graham County, Arizona. A total of 5,179 children were examined by three doctors trained in thyroidal examinations. Between 1965 and 1968, almost 11,000 examinations were performed on 4,831 subjects in grades 6 through 12. Of these, 330, or 6.8%, had some form of thyroid abnormality. Between 1969 and 1971, another 1,653 subjects in the twelfth grade were examined; 154, or 9.3%, had some type of thyroid abnormality.

After the initial screenings, a panel of experts examined these problems and reduced the total number of abnormal thyroids to 201 of 6,483—3.1%. (More than half, 104, of these abnormalities were adolescent goiters; there were two malignant neoplasms identified by the panel during this six-year period.)[30] Rallison concluded that there was "no significant increase in the occurrence of thyroid neoplasms in children in Utah and Nevada exposed to fallout radiation when compared to the control group in Arizona.[31] However, mindful of the latency period prior to the onset of thyroid cancers (as much as 25 years after exposure), he cautioned that additional studies would be needed to follow up on this epidemiological study, which was begun less than two years after the end of above-ground testing at the Nevada Test Site. (The new, major epidemiological study conducted by the University of Utah will enable Dr. Rallison to do a follow-up study in which he will re-examine 1,500 of the subjects—now adults—in Utah and Nevada, and nearly 1,500 of the subjects from the Arizona control group.)

His 1965–1970 study was admittedly flawed. The children were not asked about the contaminated milk (some of which, ironically was shipped to Las Vegas.) Also, thyroid cancer is usually not fatal, and death certificates could not be effectively used for the study.[32]

"Radioactive material that is ingested or swallowed is 'gathered and stored' in the thyroid gland," Rallison said recently. "It is possible that there was no perceptible increase in the incidence of thyroid disease in the 1965–1970 study because it was too early, that the cancers, if there were to be any, had not yet had time to develop. With this new study, we intend to locate as many as possible of the school children we examined more than fifteen years ago and re-examine them."[33] Given the poor records (nonexistent through 1958) of radioactive contamination of the milk system, Rallison has said that "an important part of this study will be the reconstruction of the milk distribution system of the area for that period."[34]

In a candid interview late in 1983, Rallison was of the opinion that there "will be no clear causality statements" emanating from the new study being directed by Dr. Joseph Lyon. Another medical person who shares similar thoughts on this issue is University of Utah President, Dr. Chase Peterson.

In 1979, Peterson, then vice president of the medical school, made the following statement when he appeared before a congressional committee that was investigating the issue of radiation and cancer: "Now, the legal proof is not a matter of our concern. The scientific proof . . . will never give 100 percent proof. And you gentlemen should not be looking for it. No doctor has ever seen a penicillin molecule fight a streptococcus germ and kill it, yet it's incontrovertible, I believe, that penicillin kills streptococcus and prevents rheumatic fever. The point is that the data is by association. I think the same will be true here. We will try to establish an association in time and place, and we hope that the data then can be reproduced."[35]

5. The Lyon Study (1979): Childhood Leukemias in Southern Utah

The questions asked by Rallison in his early study and follow-up analysis— "who was there, what they were dosed with, and what happened to them?,"[36]—are the classic epidemiological questions raised by the medical researchers in these Utah population studies. Basic questions regarding health disorders were also raised by Dr. Joseph L. Lyon in his study, "Childhood Leukemias Associated with Fallout from the Nuclear Testing." This watershed study, published by the *New England Journal of Medicine* in 1979, made inquiries about the possible association between illness, i.e., childhood leukemias, and low levels of radiation downwind of the Nevada Test Site. This study was to become "the primary study suggesting a link between the atmospheric testing program and illness in downwind residents."[37]

Dr. Lyon, a committed medical epidemiologist, occasionally becomes dramatic in his defense of qualitative studies that focus on the adverse health effects on human populations. Recently testifying before a congressional committee, he pointedly distinguished between quantitative dosimetric studies and his qualitative epidemiologic studies:

While dosimetry is important, you cannot conclude that there are no adverse health effects based on dosimetric study . . . I feel that a study of adverse health effects must be done rather than rely on imprecise and inadequate dose measurements for inferences about human health effects on fallout in this population. The appropriate method of determining health effects is *to measure the health of the population, not* to guess at the level of exposure and use uncertain extrapolations from other studies to predict the health effects that would be produced [my italics].[38]

At the same time, he accused the government of withholding the funding of these epidemiological studies and similar research efforts, in an attempt to thwart them:

The government seems to be dragging its feet in terms of setting up any human effects studies. It certainly moved substantially faster in setting up dosimetric studies. [This approach] is reminiscent of the 1950s when inadequate measurements of radiation exposure were made and the population was assured that those levels could not produce disease.[39]

Lyon, in sum, is a medical scientist who sees the value of focusing carefully on the residential histories of individuals living in the downwind areas to gauge relationships rather than relying on 30-year-old, imprecise, possibly incorrect dosimetry data generated by the Atomic Energy Commission, the Air Force, and the Public Health Service.[40]

Using epidemiological strategies in his important 1979 essay, Lyon and his associates focused on possible relationships between the low levels of radiation from the fallout and the onset of childhood leukemias. They reviewed all childhood deaths from leukemias (under 15 years of age) that occurred in Utah between 1944 and 1975, assigning these statistics to a high or low exposure cohort. Lyon made a basic assumption that Utah citizens were exposed to the fallout from at least 26 of the almost 100 above-ground atomic bombs detonated between 1951 and 1958. Lyon concluded that there was an important association between the increased risk of mortality from childhood leukemias and residence in 17 southern Utah counties (between 1951 and 1958), where what he termed the high-fallout/high-exposure group of Utahns had lived.

The number of childhood leukemia deaths occurring in the *high-fallout* counties, i.e., those closest to the Nevada Test Site, were compared within the cohort by dividing the high fallout population into *low exposure* deaths (between 1944 and 1950, and after 1958) and *high-exposure* deaths (between 1951 and 1958). Lyon then compared these results with the results generated from observing leukemia mortality in the *low-fallout* counties, which were broken down into the *low-* and *high-exposure* cohorts.[41]

The results indicated a significant increase in the number of childhood leukemia deaths in the high-fallout/high-exposure cohort. Lyon concluded that 13 deaths from leukemia would have been expected between 1944 and 1950, but that 32 occurred. Using a standardized Leukemia Mortality Ratio, the ratio of actual to expected deaths was 2.44, a twofold increase in the normal childhood leukemia death rate for the high-exposure/high-fallout group (Figure 17).[42]

Lyon concluded that it would have taken a bone marrow dosage of between 6 and 10 rads to cause the excessive number of leukemia deaths that he found in his research. Setting aside other possible explanations for the excessive childhood leukemias, Lyon's study concluded that the "excessive leukemia deaths were due to ionizing radiation exposure," with the caveat that: "the association that we report does not in itself imply an etiologic relation. The lack of accurate exposure data makes it impossible at present to determine a dose-response effect,"[43] i.e., to study a population that has been exposed to different doses of ionizing radiation and demonstrate that the risk of cancer increases with the increasing radiation dose.[44]

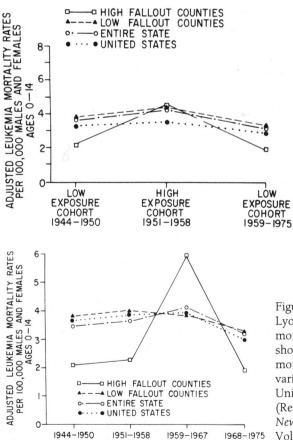

Figure 17. Two figures from Lyons' 1979 study on leukemia mortality in southern Utah, showing the variations in mortality rates from leukemia in various Utah counties and in the United States generally. (Reprinted by permission of *The New England Journal of Medicine*, Vol. 300, p. 399, 1979.)

Later that year, however, Lyon testified before the congressional committee that was examining the effects of low-level radiation. During this session, Orrin Hatch, (R-Utah) asked Lyon: "Is there a high probability that the leukemia caused has been caused from the radioactive fallout?", to which Lyon responded:

> That's the only conclusion I can draw. The more we look at the data, the more that becomes evident as the cause, and the other things we've been through additionally seem to drop by the wayside.[45]

When Hatch asked him whether it was highly probable that fallout radiation was the only cause of the high incidence of leukemia in southern Utah, Lyon's response was direct: "Radiation is the leading contender, and whatever is in second place is a long distance behind."[46]

In a 1980 study entitled "Cancer Incidence in Defined Populations,"[47] Lyon added some additional insights about the results of his Utah epidemiological study. Lyon repeated his concerns about the "uncertainties in the dosimetry" and his feeling that the AEC's recorded dosages were very "likely underestimated."[48] He also mentioned the other factors the research group had discounted in its efforts to explain the increase in childhood leukemias: there was neither industrial exposure nor excessive x-rays, particularly among the pregnant women. With the exception of chance clustering or some unknown factor, "exposure to ionizing radiation has been the one factor that has consistently been shown to increase leukemia risk," Lyon concluded in his Banbury Report essay.[49]

Land's Critiques of Lyon

Dr. Charles E. Land of the National Cancer Institute, Bethesda, Maryland, has been a persistent critic of Dr. Lyon's work. In two highly critical essays,[50] Land went into detail about the many "pitfalls" of Lyon's epidemiological research and conclusions.

Land has two basic complaints about the Lyon findings: (1) Lyon's exposure information pertains to large cohorts, not to individuals—there is no correlation with bone marrow dose in the high-exposure/high-fallout cohort; (2) the fact that there was a *decrease* in other types of cancer in the high-fallout/high-exposure group suggests another explanation for the increase in the mortality rate for childhood leukemias. "It is unlikely that radiation fallout from the Nevada weapons tests caused both an increase in leukemia mortality and a decrease in deaths from other childhood cancers."[51]

Land concluded his editorial critique with a call for additional research to "more thoroughly" address the leukemia and cancer case in terms of the history of the exposure, diagnostic information, residence, estimated radiation dose, etc. He also concluded, "it is in the nature of things for cancer mortality rates based on small populations to vary over time."[52]

Land's more recent critique of the Lyon report, published in 1984, is the more scholarly of the two rebuttals of Lyon's studies. After reviewing Lyon's data, Land and his associates at the National Cancer Institute (NCI) concluded that the atomic testing in Nevada did not cause the significant increase in childhood leukemia deaths. The Lyon results "suggest that the purported association [between radioactive fallout and increased childhood leukemia deaths] merely reflects an anomalously low leukemia rate in southern Utah during the period, 1944–1949."[53] Lyon's critical linkage "depends on the assumption that the extraordinarily low rate observed in southern Utah between 1944 and 1949 is an accurate representation of the true population risk during that time. A more likely, but nevertheless speculative suggestion, is that the early leukemia deficit was an anomaly related to underdiagnosis of leukemia or to competing mortality from other childhood disease."[54]

According to Land, Lyon had evidently ignored the fact that during this period (1944–1950) there was only one Utah Board-certified physician for a population of about 125,000 in the 17 southern counties of Utah, and that, given this gross lack of medical staffing, misdiagnosis and nontreatment (due to "the difficulties of transportation in areas of low population density")[55] should be expected. "It is the experience before 1950, rather than subsequently, that requires explanation," stated Land.[56] (A question unasked by Land was why these pre–1951 conditions reappeared *after* 1959.)

After presenting National Center for Health Statistics (NCHS) demographics and using eastern Oregon as a control group, Land concluded that (1) the southern Utah leukemias mortality ratio in the high-exposure/high-fallout group was *less* than in Oregon,[57] and (2) that there was no statistically significant difference in standardized mortality ratios between northern and southern Utah.[58] "The evidence for an increase in childhood leukemia mortality in southern Utah as a result of exposure to radioactive fallout between 1950 and 1958 appears, on closer examination of available data, to be slight or nonexistent,"[59] he concluded.

6. The Caldwell et al. Study (1980): Leukemia in Military Personnel at the Nevada Test Site

Dr. Glyn Caldwell and his associates at the Center for Disease Control (CDC) in Atlanta, Georgia, focused on the August 31, 1957 "Smoky" shot at the Nevada Test Site after they learned of the diagnosis of acute myelocytic leukemia in an army veteran exposed to ionizing radiation during the nuclear test. In 1976, the Center for Disease Control began a leukemia study of military servicemen who had participated in military maneuvers at the Nevada Test Site in close proximity to the Smoky blast. As a result of Paul

Cooper's request for information about whether the "Smoky" radiation could have caused his terminal leukemia, an epidemiological study was undertaken to measure the frequency of leukemia in these military personnel.

Paul Cooper was one of 250 military personnel who went within one hundred yards of ground zero, minutes after the Smoky blast. "The heat was so intense," he said, "I felt as though my uniform was going to catch on fire."[60] He did not know about radiation contamination in the area although "scientists who approached from a different direction had to abandon their attempt to retrieve instruments because of intense radioactivity. The instruments were left in the field for several days until the radiation dissipated."[61]

Once the Center for Disease Control found out about Cooper's leukemia and his association with the atomic testing at the Nevada Test Site, Dr. Glyn Caldwell launched the investigation to determine if there were leukemia clusters among the former Smoky soldiers. Caldwell knew the task would be enormous, but he was determined to investigate any carcinogen that caused cancer. "I was very interested in this study and pigheaded enough to see it through," said Caldwell.[62]

The Center for Disease Control researchers identified most of the participants who had been at the shot, reconstructed the radiation dose each person received, and related the leukemia and cancer incidence in this cohort to the dose information and to expected levels of incidence of leukemia and cancer in nonexposed control populations.[63]

The results were striking: there were 9 Smoky leukemia deaths where only 3.5 were expected. "In no case did we find a strong history to suggest an exogenous cause for leukemia,"[64] Caldwell concluded. The high leukemia mortality rate suggested to the medical researchers "either a greater dose than estimated (the examination of film badge readings, worn by some of the soldiers during the blast, showed very low doses for several leukemia victims) or a greater effect per rad at low-dose levels."[65] Contrary to the 1980 Shaeffer Report's review of this study, which stated that the "CDC has not related the increased incidence of leukemia to radiation dose," the Caldwell study attributed the excess leukemia to the radiation exposure.[66]

Bond and Hamilton's Critique of the 1980 Caldwell Study

V.P. Bond and L.D. Hamilton, medical researchers at the federally run Brookhaven National Laboratory, were extremely critical of the first Caldwell study. More remarkable to them than Caldwell's results was the "extent to which the authors discuss difficulties in reaching conclusions from their data."[67]

Their argument with Caldwell was that the Center for Disease Control

Smoky study had only looked at a little more than 3,000 persons exposed to less than 1 rad of radiation (according to the extremely questionable Atomic Energy Commission data). Since it is "generally agreed" that epidemiologists must work with large study populations, "studies with small doses and small populations can shed no clear light on the risk of cancer induction by low-level radiation."[68] In a sharply worded conclusion, the Brookhaven scientists wrote that the Caldwell study "perpetuates a fallacious reasoning: that several separate studies, each inadequate to demonstrate quantitative relationships between dose and response, should somehow override the few large, statistically valid studies from which risk estimates are now derived."[69]

Lyon's Response to the Bond/Hamilton Criticism

Since Bond and Hamilton were also (indirectly) critical of the Lyon study of childhood leukemias in southern Utah, Dr. Joseph Lyon and one of his associates responded to them.[70] For Lyon, the Caldwell study provided additional evidence that fallout from the atomic testing at the Nevada Test Site increased the risk of cancer. Caldwell's estimate that a "radiation dosage of about 70 rads would be necessary to produce the carcinogenic effect found, but available film badges showed dosages of only 0.5 to 1.0 rads," was an important observation.[71] The basic inconsistency could only be explained in one of a few ways: (1) the study results were due to chance; (2) "low levels of radiation may cause cancer more frequently than thought"; or (3) "the radiation received was much greater than originally estimated by the crude dosimetry available."[72] The Lyon study encountered the same type of dosimetry problem during the research. Bond and Hamilton, however, "glossed over" these issues.

Warming to the task of criticizing the government's "hired guns," Lyon argued that Bond and Hamilton's reasoning about small numbers in a population was itself fallacious. "The question of statistical power is irrelevant when a study has already shown an excess of cancer, and no causal association can be inferred without a statistical association between the events. The assumption that small studies produce significant results only by chance is simply not the case."[73]

Lyon concluded by pointing out that "the facts in the case of exposure of human populations to nuclear fallout are that all of the studies to date have found excess leukemia in their exposed populations. Since these three small studies (southern Utah, Marshall Islands, "Smoky" test participants) all point in the same direction, it seems warranted to look at larger populations for evidence of adverse health effects owing to fallout, rather than to question the motives and credibility of investigators who engage in research on these populations."[74]

7. The Caldwell et al. Study (1983): Leukemia in Military Personnel at the Nevada Test Site

In a 1983 follow-up to their study of the adverse effects of the Smoky shot on the military participants at the Nevada Test Site, Caldwell and his associates, now extremely sensitive to the scientific battle raging around them, investigated the health status of almost 96% of the military personnel (3,072 of the 3,217) who participated in the military exercise.[75] They found, after contacting these personnel, reviewing health records and "the incomplete exposure data,"[76] and calculating the expected incidence and mortality figures for leukemias and other neoplastic diseases, that there were 112 new cancers observed whereas the expected figure was 117, and 64 deaths were attributed to cancers and leukemia, whereas the expected figure was 64.3.[77] There was, however, a statistically significant increase in the frequency of occurrence and mortality for leukemia: 10 cases were observed when only 4.0 were expected, and 8 deaths occurred, when only 3.1 were expected.[78]

Their conclusion was a tenuous one. After 22 years, there has been no significant increase in the incidence of mortality from any cancer other than leukemia. Consequently, the 1983 Caldwell conclusion was much more tentative than the initial report. It was that the leukemia findings were either due to chance, factors other than radiation, or to a combination of factors including the radiation these persons were exposed to during the Smoky test of August, 1957.[79]

Reading the report's conclusion, one is struck by the medical researchers' extreme sensitivity to the debates over this issue. "We expected that some sort of accident was a most likely [explanation]. We did not expect to find only the small exposures recorded on the firm badges or *to become embroiled in the low-dose controversy* [my italics] . . . However, the low, overall cancer incidence and mortality, the low level of exposure (uncertain as it is), suggest that this one positive finding—leukemia deaths—may be either attributable to chance or the result of an unknown combination of factors. Furthermore, this conclusion cannot be generalized to include participants at other nuclear tests or resolve the low-dose controversy."[80]

Caldwell's 1977 "pigheadedness" evidently ran into intense pressure from government medical research critics such as Lamb, Bond, Hamilton, and others. Caldwell's sensitivity to the "low dose controversy" that swirled around these kinds of studies is understandable. Credible medical epidemiological studies that draw associations between increased incidence of cancers and exposure to fallout from the atomic testing exercises may place the federal government in (tort liability) jeopardy; beyond the medical disagreements, there is the fundamental liability question that these medical studies address.

8. The Beck and Krey Study (1983): Radiation Exposures in Utah

Harold L. Beck and Philip W. Krey, of the U.S. Government, recently published an essay on the subject of the impact of the fallout from the tests on the downwind residents of Utah.[81]

Beginning their review of the Lyon study in the same manner as Bond and Hamilton and Land did, these researchers questioned the reliability of the Lyon results about linkages between fallout and increased incidence of leukemias. For the leukemias to have resulted from the fallout, they argued, "based on currently accepted estimates of risk," significantly higher exposures to radiation than previously reported would have been required.[82]

Beck and Krey then attempted to quantitatively reconstruct the radiation exposure received by the Utah population during the above-ground testing by taking soil measurements for cesium–137 from 150 sites across the state to determine the external radiation exposure and bone dose.

Using this radiobiologic method, they found that levels in northern Utah exceeded their expectations. "This resulted in an estimated mean population exposure for low-fallout counties approximately 50% greater than for the group of high-fallout counties."[83] Based on this gross measurement, they concluded that residents of Salt Lake City had received greater radiation exposures than most Utah residents living closer to the Nevada Test Site— except for residents of Washington County, Utah[84] (the county closest to the Nevada Test Site).

Based on the government's "accepted best estimate of lifetime risk,"[85] Beck and Krey calculated that the doses received by the southern Utahns were much lower than those required to cause the significant increase in childhood leukemias that Lyon had discovered. Whereas Lyon indicated that there had to be a bone marrow dosage of at least 6 to 10 rads to produce the excess of childhood leukemias, Beck and Krey's study concluded that "the population of the high fallout region received an average bone dose of about 0.3 rads."[86] Their final comments rejected the Lyon study. "It seems unlikely that the excess leukemias observed by Lyon resulted from exposure to the NTS fallout."[87] They did not, however, pose any alternative explanation for the excess of childhood leukemias in the high-fallout area during the high exposure period (1951–1958).

9. The Johnson Study (1984): Cancer Incidence in Utah

Dr. Carl J. Johnson, a physician who regularly participates in these controversial testing issues (in Utah and in Rocky Flats, Colorado) conducted a study that was published in the *Journal of the American Medical Association* in January, 1984, which examined the incidence of cancer in the area of heavy radiation fallout downwind from the Nevada Test Site.[88] Noting the rising incidence of cancers in populations that had been exposed to radiation fallout (Hiroshima and Nagasaki residents and the Marshall Islanders) and also noting that Dr. Robert E. Alexander, Office of Standards Development, U.S. Nuclear Regulatory Commission, indicated that the worldwide effects of nuclear weapons testing may cause 29,000 to 72,000 deaths from cancer, and 168,000 adverse genetic effects (all generations), Johnson focused his epidemiological research on the "health effects in local civilian populations downwind from nuclear sites and subject to greater exposures of fallout."[89]

Johnson measured cancer incidence in a 1951 cohort of Mormon families (totaling 4,125 people) in southern Utah near the Nevada Test Site who could still be located in 1981. He then compared the results of his measurements with the information he had gathered about Utah Mormons across the state (Table 4). His study group consisted of people living in towns with *heavy fallout exposure* such as St. George, Parowan, Paragonah, and Kanab in Utah; Fredonia, Arizona; and Bunkerville, Nevada.

Trained volunteers took a survey of this population between April and December, 1981. The epidemiological questions related to family history; medical consequences of exposure to fallout; smoking experience; employment; and diagnosis of cancer.[90] Supplemental information was taken from those who had a medical diagnosis of cancer. After generating this study-group information, Johnson compared his figures with the figures on the control group (Utah Mormons statewide).

Johnson's results showed that, as compared with the control group's mortality records, there was an excess of cancers and leukemias in the study group:

> There was an excess of 109 cases of cancer (288 cases observed; 179 expected) in this southwestern Utah population of 4,125 . . . Leukemia was preponderant early and persisted later, compatible with a prolonged period of exposure to radioactive fallout during 1951 through 1962 . . . Cancer of the thyroid gland was prominent in the exposed group.[91]

His conclusion was clear:

> The comparison provides assurance that the excess incidence of cancer is actually caused by the exposures to radioactive fallout. No other explanations for these effects were discovered in the investigation, i.e., smoking, occupational history, or industrial point sources of carcinogens.[92]

Ominously, he ended his study with these words:

> A burden of radiation-induced cancer throughout the state can be expected, because an excess of childhood leukemia has been reported for the entire state, and this observation is an early warning of other classes of radiation-induced cancer to appear later.[93]

Table 4. Cancer incidence in a Mormon population residing in southwestern Utah exposed to radioactive fallout compared with cancer incidence for all Utah Mormons[a] (Reprinted by permission of the *Journal of the American Medical Association*, January 13, 1984, Vol. 251, p. 232, copyright 1984, American Medical Association.)

Type[b]	Class (International Classification of Disease No., Revision 8)	Utah Mormons 1967–1975, Rate[c]	High-Fallout Area[11]						Fallout-Effects Group					
			1958–1966			1972–1980			1958–1968			1972–1980		
			Observed[d]	Expected[d]	Rate	Observed	Expected	Rate	Observed	Expected	Rate	Observed	Expected	Rate
A + B	All sites (140–207)	228	118[f]	76.0	354	170[f]	102.8	377	31[f]	5.3	1,322	33[f]	7.1	1,064
A: Cancer of more radio-sensitive organs	Lungs, respiratory (162, 160, 163)	16.5	7	5.8	20.0	6	7.7	12.8	1	0.8	21	1	0.8	22
	Upper GI tract (141–150)	4.4	4	1.5	11.4	4	2.0	8.9	0	—	—	1	0.2	26
	Stomach (151)	7.0	9[f]	1.8	34.5	5	2.8	12.5	3[f]	0.2	138	2	0.4	36
	Colon (153)	19.7	5	7.0	14.0	14	8.1	34.1	2	0.6	66	3	0.7	87
	Breast (174)	31.9	8	8.8	29.1	27[f]	14.2	60.7	2	0.7	96	8[f]	0.7	347
	Thyroid (193)	3.7	6[f]	1.4	16.3	14[f]	1.7	30.8	0	—	—	1	0.1	26
	Leukemia	7.9	19[f]	3.6	41.7	12[f]	3.4	28.1	9[f]	0.2	307	1	0.1	70
	Lymphoma	10.7	4	4.2	10.1	10	5.2	20.7	1	0.2	48	5[f]	0.3	199
	Total	101.8	62[f]	34.1	177.1	92[f]	45.1	208.6	18[f]	2.7	676	22[f]	3.3	813

B: Other cancer

Melanoma (172)	5.7	9.0	3	1.9	195	9	2.6	85	2[f]	0.1	13	1	0.4
Brain, CNS (191, 192)	4.7	15.0	5[g]	1.6	82	4	2.5	131	2[f]	0.1	70	1	0.1
Bone, joint (170)	0.8	8.3	3[f]	0.3	101	5[f]	0.4	—	0	—	—	0	0.1
Other	115.0	144.6	45	55.8	1308	50	52.8	430	9[f]	2.4	168	9	6.2
Total	126.2	176.9	56[g]	59.6	1684	78	58.1	646	13[f]	2.6	251	11	6.8
A/B	0.807	1.001	—	1.239	—	—	1.046	—	—	3.239	—	—
$\left(\dfrac{\text{A/B (subpopulation)}}{\text{A/B (Utah)}} - 1\right) \times 100$	—	+24%	—	−53.5%	—	−29.6%	—	—	+301.4%	—	—

a Radioactive fallout occurring from atmospheric detonations of nuclear weapons at the Nevada Test Site between 1951 and 1962. Data have been adjusted for age and sex; age adjustment for the three populations by the direct method with the 1970 US white population as the standard (JNCI 1980;65:1169).

b "A" Classes of cancer (of the more radiosensitive organs) are those found in excess among the survivors of the nuclear bomb detonations at Hiroshima and Nagasaki: leukemia, lymphoma, cancer of thyroid gland, lung, esophagus (included here are other cancers of the upper gastrointestinal [GI] tract 141-150), stomach, colon, and breast.

c All rates are average annual age-adjusted incidence of cancer per 100,000.

d The number of cases observed is compared with the number expected. Expected case numbers were calculated by the following approach: the number of cases in each class times the average annual age-adjusted cancer incidence rate for Utah Mormons divided by the average age-adjusted cancer incidence rate for the study population (Environ Res 1981; 25:86).

e High-fallout areas consist of the St. George area, Parowan, Paragonah, and Kanab, Utah; Fredonia, Ariz; and Bunkerville, Nev. For some cases, dates of diagnosis of cancer were not recalled with certainty. These were allocated to the broad time periods indicated in the table above. Those not clearly falling in a time period were assigned to a time period in the sequence in which they appeared, with conservative effect, ie, assigned equally to the early (1958 through 1966), interim (1967 through 1971), and intermediate (1972 through 1980) time periods. The interim period was not studied because the design is to compare an early period with a later period.

f Indicates significant at P = .01.

g Indicates significant at P = .05 (two-tailed test).

10. The Lyon Study (1982-   ): Leukemia in Southern Utah

In 1982, scientists at the University of Utah Medical Center began a major, six-year epidemiological study to determine if an excess of thyroid cancer and childhood leukemia has occurred in the region, and if an association exists between their incidence and the radioactive fallout. The study is led by Dr. Joseph L. Lyon, Associate Professor of Family and Community Medicine and author of a major study showing strong connections between childhood leukemia in southern Utah and the above-ground production of low-level fallout.[94]

Sponsored by the National Cancer Institute (NCI) at a cost of $6.5 million, this major study is attempting to determine the amount of radiation that southern Utah received and to determine the leukemia and thyroid cancer rates for the local populations. They then will 'overlay' the radiation dose information on the cancer rates to see if the cancer rates are higher in the areas of high radioactive fallout exposure and will compare the cancer rates of the exposed populations with those of large control groups in Idaho and Arizona. "The control groups consist of individuals of like genetic backgrounds and lifestyles" who were not exposed to radiation from the atomic bomb tests.[95]

A more quantitative radiobiological experiment will also be conducted for the U.S. Department of Energy. This radiation dosimetry effort, performed by Dr. McDonald E. Wrenn, director of the Division of Radiobiology at the University of Utah Medical Center, and Dr. Charles W. Mays, professor of pharmacology, will employ a new technique called *thermoluminescence*, which enables researchers to determine, from quartz found in building materials such as bricks, the amount of radiation fallout a building has received since its construction.[96]

Summary of the Studies on Possible Adverse Health Consequences in the Downwind Population

At least two-thirds of the epidemiological studies on cancer incidence and its possible association with fallout would agree with Dr. McDonald E. Wrenn that, for the people in the downwind towns of southern Utah and Nevada, the "fallout from the NTS was the clearly dominant source of exposure."[97] Six of the nine studies, the Weiss, Knapp, Center for Disease Control, Lyon, Caldwell I, and the Johnson studies go beyond Wrenn's cautionary language and state that a fairly clear association exists between the downwinders' exposure to radioactive fallout and the increased incidence of leukemia.

Five of the six studies clearly focus on leukemia because of the short time that elapses between exposure to a carcinogen—radioactive fallout—and the onset of leukemia, whereas solid mass cancers take 25 to 50 years to appear. Two of the three studies that found no association between the fallout and increased cancers in the population—Rallison and Caldwell II—cautioned their readers that perhaps not enough time had elapsed for them to find any increases in solid tumor cancers in the Utah and Nevada cohorts. Indeed, Rallison has set out to reinvestigate the children he interviewed fifteen years ago in whom he found only a small number of thyroid abnormalities.

In a study of the mortality of the atom bomb survivors,[98] Hiroo Kato, William Schull, and their colleagues found that among the Hiroshima and Nagasaki survivors leukemia mortality figures peaked in 1950–1954, and have since continued to decline. However, solid tumor cancers have increased as the cohort has aged. The late somatic effects of atomic bomb radiation seem to be limited to an increase in malignant tumors. Although cancer morbidity and mortality has increased since 1960, for the first time, in 1980, the increase is statistically significant (in Nagasaki).[99] There is no increase in mortality figures other than solid cancers, and this factor is now significantly associated with the radiation exposure from the 1945 bomb.[100]

Eight of the nine studies either draw a fairly strong association between the fallout from the atomic tests and increases in leukemias and cancer or suggest that it is still too early to make a strong association between the fallout and the incidence of cancer.

Only the critics, Land and Caldwell, and Beck and Krey, all of whom are employed by the federal government, discount the associations. The battle line in the "low dose controversy" is clearly drawn along the government/non-government axis. "Scientists defending the NTS testing activities work for the government," Lyon commented in a recent interview. "What would be . . . their credibility if the NTS were run by the Dow Chemical Company rather than the government? . . . Unfortunately, that's not the case with the NTS. The federal government runs it, and the scientists they marshall to argue for it are credible, reputable people in many instances, although some are 'hired guns' who work over critics pretty heavily—and we've been worked over pretty heavily."[101]

In addition, there is also a philosophic difference of opinion between the radiobiologists doing dosimetry reconstruction research and the medical epidemiologists who are trying to reconstruct the human and environmental habitat in an effort to find the answer to the riddle of fallout and cancer increases in the downwind communities. Essentially, the radiobiologists take a predetermined causal relationship and try to discover how medical science can eliminate risk by setting up preventive standards for the community to follow. Presuming a causal relationship, they measure precise tissue doses, etc., and then develop standard risk estimates based on known dose relationships.[102]

The epidemiologists, on the other hand, try to develop associations between adverse health effects and possible carcinogenic agents that may be the cause. What does an epidemiologist need in order to be able to say that there is a relation between radiation exposure and disease? "Epidemiology has been able to demonstrate what most people would accept as causal relationships without getting down to precise tissue dosages," said Dr. Lyon.[103] The insistence on needing specific dosage figures before causality can be determined is grossly unfair for two reasons, Lyon claims. First, "you don't do that for cigarette smoking, asbestos, for a variety of chemicals in the work place. Second, "trying to find the dose effects 30 years after the event is not an easy task. It certainly would have been made a lot easier if we had set up an ongoing health study in 1952 or 1953. Had we had careful monitoring, many problems we face today wouldn't have been there." Lyon added cheerily, "you have to study what you have!"[104]

The "folklore"of the downwinders—that there were much higher doses of radioactive fallout than the Atomic Energy Commission recorded—may not be far from reality. A number of the researchers concluded that either the dosimetry readings were incorrect or that the consequences of human exposure to low doses of radiation are more insidious than they had imagined.[105]

A critic of the "associational"medical epidemiologists has said that, "in this area, public policy decisions need to be made in the absence of scientific evidence."[106] However, based on a review of the research to date and on current research projects, there is a growing body of scientific evidence that supports the association between exposure to radioactive fallout and increasing incidence of leukemia and cancer in the downwind community.

Public policy makers, federal judges, national legislators, as well as federal bureaucrats and members of the executive branch, can draw upon this growing body of medical research that is slowly accumulating the data needed to demonstrate the association between these two factors, in a specific time and place, with a defined group and geographic setting. While the medical researchers have been caught up in the controversy, their studies speak for themselves.

Janet Gordon's complaint that the medical researchers "are not looking at us" as they go about their radiobiological and epidemiological studies is not quite true. While some medical researchers are doing appropriate radiobiological work with bricks and small squares of earth samples, there are others, such as Dr. Lyon and his colleagues, who are, through planned University of Utah Survey Research Center telephone surveys, returning to the community once again, to "speak" with the people and review the biological and environmental area where these events occurred twenty to thirty years ago.

The current epidemiological associations are also the kind of "evidence" the plaintiffs had to develop as they made federal court appearances in Salt Lake City, seeking a legal remedy for the injuries and deaths of members of their families due to the negligence of federal government agents who implemented the Atomic Energy Commission's nuclear testing program at the Nevada Test Site.

In the medical controversy, as in the legal and political debates surrounding the issue of radiation and increases in cancer, Lyon's comment about the role of the federal government is valuable: "There is a very definite government position that doesn't seem to talk science without being able to talk compensation simultaneously, and you have problems sorting out the scientific, objective research from the compensation problem."[107] His observation is an understatement of the justice dilemma that has unfolded in the federal district court in Salt Lake City

6 The Downwinders Discover the AEC Records and Seek a Remedy for Radiation-Caused Injuries in Federal Court, 1978–1979

These cancers are torts, injuries inflicted by one person on another.
Note, *Yale Law Journal*, 1981

The idea that there can be a wrong without a remedy is anathema to our sense of justice.
Bruce Clark, *Air Force Law Review*, 1974

Making the Information Available to the Litigants

After the leukemia and cancer began striking the downwinders in the early 1960s, they suffered in isolation for ten to fifteen years. As a community they drew even closer together, giving each other solace and sympathy, but they did little to make their tragedy a national issue. Their sole, early effort to tie the fallout to its consequences had been the unsuccessful suit attempting to link the sheep's deaths to the low-dose radiation that had blanketed their farms in the early days of the nuclear tests.[1] It was not until the late 1970s that a number of separate yet related events led these people living in the small towns east of the Nevada Test Site to conclude that they had been physically, legally, and morally wronged by agents and agencies of the federal government, and to seek recompense, in the form of monetary damages, in the federal courts.

Between 1978 and 1979, local leaders galvanized the residents of the downwind communities and, by the summer of 1979, they instituted a lawsuit in U.S. Federal District Court in Salt Lake City, charging the federal government, specifically the Atomic Energy Commission operators at the Nevada Test Site, with gross negligence and carelessness in the conduct of the nuclear testing program.

By 1978, the scientific community, politicians, and downwind residents were aware of both the Lyon study, involving the leukemia incidence in southern Utah, and preliminary assessments of the Smoky leukemia study.

In addition, information about the early medical studies, which the Atomic Energy Commission previously had repressed (the Knapp, Weiss, and the Gofman and Tamplin reports), was reluctantly released by the federal agencies. These disclosures were made because news reporters requested the information through the Freedom of Information Act after Lyon's report was released.

In a major story published January 8, 1979, in the *Washington Post*, for example, Bill Curry revealed that "federal health officials had evidence as early as 1965 that excessive leukemia deaths were occurring among Utah residents exposed to radioactive fallout from the U.S. atomic bomb tests."[2]

Legislative Action

Because of the emergence of the medical reports linking cancers to the nuclear testing, congressional action began in 1978. In January, 1978, Representative Lee Carter (D-Tennessee), through his House Subcommittee on Health and the Environment, began the initial set of hearings on the relationship between the downwinders' health and the effects of low-level radiation exposure from the nuclear tests. His hearings were a milestone: "for the first time, government officials, representing both the military and the AEC, admitted that errors had been made during the atmospheric testing period. Their testimony alerted the public to the possibility of government negligence, and placed the issue on the government's active agenda at a time when it could not be covered up."[3]

Between 1979 and 1981, congressional committee hearings revealed for the first time the activities of the federal government and its operatives (especially the AEC and Public Health Service agents) as they implemented the continental atmospheric testing of nuclear weapons with little demonstrated regard for the mounting evidence that the health of American citizens was being jeopardized by the tests.[4]

A 1980 report prepared for the House Committee on Interstate and Foreign Commerce, entitled "The Forgotten Guinea Pigs," sharply condemned the federal government and summarized the feelings of the legislators involved in all of the congressional hearings on the subject. The report concluded with the following comments:

1. Despite known information about the hazard of radiation exposure, the government "failed to give adequate warning to the residents living downwind from the test site regarding the dangers posed by the radioactive fallout emitted during the atmospheric nuclear test operations."
2. "The radiation monitoring system established by the government during the atmospheric nuclear testing program was deficient in giving accurate estimates of radiation exposure."

3. The government "falsely interpreted and reported radiation exposure rates so as to give an inaccurate estimate of the hazards."
4. The government "knowingly disregarded evidence which questioned the accuracy of the government's measurements of radioactivity emitted from the test site as well as the adequacy of the then-employed-radiological safety standards."
5. Exposure to the radioactive fallout "was, more likely than not, responsible for the serious adverse health effects suffered by the downwind residents."[5]

Executive Actions

In May, 1978, after Representative Lee Carter's House Subcommittee hearings had concluded, President Jimmy Carter appointed his own federal interagency task force, under the direction of Peter Libassi, then general counsel for the U.S. Department of Health, Education and Welfare, to study the effects of low-level radiation on the health of the communities downwind from the testing site. (The Libassi Report, presented to the President in July, 1979, concluded that "radiation may cause irreparable change in the cells, resulting in cancer, developmental abnormalities and genetic damage.")[6]

Utah's Governor Scott Matheson spoke to Joseph Califano, Secretary of the Department of Health, Education and Welfare, and to President Carter in November, 1978. During a visit to Salt Lake City on November 27, 1978, President Carter, in response to the governor's urging, issued an order to Califano's Department of Health, Education and Welfare (1) to re-evaluate the findings of earlier medical studies on fallout's effects on human health; (2) to determine whether there was an association between leukemia and fallout; (3) to reopen the thyroid studies that had been discontinued by the federal agency; and (4) to develop a more complete study on the association between fallout and cancer.[7] (At the same time, Congress had passed Public Law 95–622, which appropriated funds for an extensive research program to be conducted by the Department of Health, Education and Welfare on the health effects of low-level radiation.)

In early 1979, Califano's agency conducted an extensive document search that produced thousands of pages of documents. (Matheson, in early January, 1979, had written letters to James R. Schlesinger, Secretary of Energy; Harold Brown, Secretary of Defense; Douglas Costle, the Environmental Protection Agency Administrator; Bob Bergland, Secretary of Agriculture; and Major General Joseph K. Bratton, Director, Military Application, Department of Energy, requesting them to assist and cooperate with Utah in "connection with the current re-evaluation of the public health impact of atomic testing conducted at the NTS.")[8]

Recalling these Atomic Energy Commission data collection and cancer research efforts, Secretary Califano's diary clearly suggests fundamentally different agendas for the federal agencies involved. A number of federal agencies, including the Department of Health, Education and Welfare, the Environmental Protection Agency, National Cancer Institutes, the Center for Disease Control, and the Department of Defense, were examining the association between fallout and the onset of cancer and leukemia. By the spring of 1979, the Libassi committee's findings were raising serious questions about governmental negligence in Califano's mind—yet his concerns were not shared by the rest of the federal establishment reviewing the controversy. As Califano has written, "We [in HEW] were deeply concerned about radiation effects and by the lack of coordination on this subject within the government. The Nuclear Regulatory Commission [the successor agency to the Atomic Energy Commission] was going its own way, trying to ignore our concerns at HEW and those of Doug Costle at EPA."[9]

At the same time, the documents started flowing out of Washington, D.C. This information became available to the public through the news reports and congressional hearings that occurred later in 1979. "Public disclosure of all pertinent documents," said Matheson before the congressional committee in the spring of 1979, "will prove valuable to scientists in their thorough and exhaustive health study. And disclosure will help the government regain public trust and confidence."[10] These documents also went a great distance in clarifying questions about the conduct of the AEC policy-makers in Washington, D.C. and, more relevant to the pending federal court litigation, the activities of the local operators at the test site.

Califano was the major catalyst, according to the Utah governor, in the move to open the files. In the fall of 1978, Matheson's anger had been aroused by a local television broadcast that dealt with the fallout issue.[11] If Matheson had not contacted the Secretary of Health, Education and Welfare while he was visiting Carter at the White House, the Atomic Energy Commission and other governmental files, long closed, might not have been opened. Matheson first met with Jack Watson, President Carter's intergovernmental relations liaison, but their discussion about providing federal assistance to those injured by the atomic testing was unproductive. Following his meeting with Watson, as Matheson was preparing to leave the White House, his press secretary arranged for him to meet with Califano (who was returning from Europe) the following morning.

Matheson left a sheaf of fallout documents with Califano's secretary and went back to the hotel, with no great expectations for his upcoming meeting with the federal official. The HEW Secretary, however, had read all the materials on the fallout issue by the time Matheson came to see him at 8:00 the next morning! Califano was extremely moved by the report and asked what the Carter administration could do to help. Matheson seized the opportunity and urged: "We need to get to the AEC documents; help us to open them up for the people to understand what really did happen at the NTS." Califano immediately agreed to speak to the President about this matter and, shortly thereafter, the President made his announcement during the trip to Salt Lake City in late November, 1978.

A very committed state executive, Governor Scott M. Matheson, and an equally supportive federal official, Joseph A. Califano, Secretary of Health, Education, and Welfare, played critical roles at this moment in the legal-political history of the fallout issue. Because of their desire for simple justice for people who alleged that their government had lied to them and failed to protect them from the dangers of nuclear fallout, millions of words were released from the federal bureaucratic files. The attorneys for the plaintiffs in the nuclear testing litigation were able to continue with the important process of discovering, through information revealed in the agency documents themselves, lines of legal strategy for the battle against the federal government.

The documents revealed to the plaintiffs' attorneys the governmental lack of due care in its implementation of the policy judgment to test nuclear weapons on the American mainland. As one of the attorneys wrote to Stewart Udall, the lead attorney for the downwind plaintiffs: "Plaintiffs in this case had the right to expect rigorous governmental efforts to protect them against the effects of fallout. What they received were assurances from their protector that all was well, when in fact, the reverse was true."[12]

The Legal Dilemma

The downwind residents had the beginnings of a definitive case concerning governmental actions at the Nevada Test Site when the Atomic Energy Commission's activities at the Nevada Test Site were displayed for public view through the medical studies, news reports, congressional committee investigations, and executive actions. The legal obstacles encountered by the downwinders were: (1) the state of federal tort law and (2) the limited ability of a federal judge to provide remedies for these persons, given the way in which precedent had developed in the applicaton of the federal tort law, the Federal Tort Claims Act (FTCA) of 1946.

The central pupose of the common law of torts is to compensate victims of injury due to negligence; to "achieve justice between individuals where one had harmed the other."[13] Under the existing law of torts, and its precedents, the injured person—the plaintiff—must show causation-in-fact, that is, the injured person must prove that the defendant's conduct caused the harm. In recent years, in addition to the "individualized, immediate wrongs for which tort law developed,"[14] such as injuries that result when a car strikes someone or when someone punches another person in the nose, a new type of injury has emerged. The "fundamentally different"[15] toxic injury occurs after time has elapsed, and is the result of "biological causation,"[16] rather than direct causation. It has been defined as injury that occurs when "a force or substance affects the human body, with the injury or disease manifesting itself months or years later."[17]

After "picking himself or herself 'out of the crowd'"[18] of downwinders who were exposed to radiation from the above-ground atomic tests at the Nevada Test Site, the biologically injured (downwind plaintiff) must recover from the government who allegedly committed the harm. The downwinders, therefore, immediately ran into the major difficulty that nontraditional tort suit plaintiffs confront: the rule that in a tort suit "the plaintiff must show that the defendant's conduct contributed to his or her injury."[19]

Before filing their legal briefs and confronting these legal dilemmas, however, the lawyers for the downwinders were required to exhaust administrative remedies. In late December, 1979, Stewart Udall filed 100 claims totaling over 100 million dollars with the U.S. Department of Energy, charging the federal government with negligent conduct that led to injury and death of persons downwind of the test site.

The Department of Energy had six months to process the claims. If the claims were denied, then the downwinders would have recourse in filing a tort liability suit in Federal District Court. Udall's view was that the United States should pay the claims of these cancer victims in the fallout areas "because these people were war casualties." While stating his sincere hope that the Department of Energy "will see the justice of our case and not force this to court,"[20] Udall and his staff continued to prepare for the litigation they felt they would have to institute after the six-month, administrative appeal period expired.

They were right. The Department of Energy did not act on the claims within the legal required time and, in August, 1979, the downwinders and their lawyers went into federal district court to seek a hearing on their contention that the government had been guilty of negligent, tortious conduct.

The Federal Tort Claims Act

The Federal Tort Claims Act (FTCA), Title VI of the Legislative Reorganization Act of 1946, was a somewhat unsuccessful effort by the national legislature to negate the doctrine of sovereign immunity.[21] Until its passage, no legal action of a civil nature could be taken against the United States government unless the government gave its permission to be sued. In 1946, the government gave its general but limited consent to be sued in tort actions in federal court. The legislative objectives were two: (1) to relieve Congress of the burden of private relief bills, and (2) to provide justice to those who had suffered injury or losses through the negligent acts of the federal government's employees.[22]

The FTCA was the "offspring of a feeling that the government should assume the obligation to pay damages for the misfeasance of employees in carrying out its work."[23] As we shall see, it falls far short of the common law principle that the tort of the servant is the tort of the master.

The act stipulates that a federal district court judge, sitting without a jury,

> shall render judgment on any claim against the United States on account of damages and/or loss of property and/or personal injury or death caused by the negligent or wrongful act or omission of any employee of government while acting within the scope of his office of employment, under circumstances where the United States Government, if a private person, would be liable to the claimant for such damage loss, injury or death in accordance with the law of the place where the act or omission took place.

There is a heavy burden of proof placed on the plaintiffs in tort suits against the defendant United States Government. The cause of action for negligence must include *five primary elements:*[24]

1. The government must have a *legal duty* or obligation imposed by law to act in a certain way.
2. There must be shown a breach of duty, failure to so act, i.e.,unreasonable, negligent (tortious) conduct by agents of the government.
3. Governmental conduct, which was negligent, must be shown to be the proximate or legal cause of the plaintiff's actual injury.
4. The governmental agents who acted unreasonably or negligently must be shown to have been under a *legal obligation* to act for the plaintiff's benefit but failed to so act.
5. Plaintiff must show actual injury.

In the atomic testing litigation, according to tort law, the downwinders had to show (1) that there was negligence by an agent or agents of the government; (2) that the government's negligence occurred during the *operational* rather than the planning stage; and (3) that the illness was caused by the continual exposure to radiation brought on by the detonations during the testing period (1951 to 1962). The Federal Tort Claims Act and the federal courts require evidence that the government's action "more likely than not" was the direct cause of the plaintiff's injury, illness or death. On the other hand, the government's defense in tort suits in federal district court proceedings, is (1) to show that the litigation began after the statute of limitations clock had run down; (2) to show that the plaintiff failed to satisfy any one of the five necessary conditions; and (3) to argue that the action challenged by the plaintiffs is a governmental action that falls under what is called the "discretionary function" exception.

The basic effect of the act is "to waive immunity from recognized causes of action and . . . not to visit the government with novel and unprecedented liabilities," ruled the U.S. Supreme Court majority in *Feres*. In the original legislation, there were thirteen exceptions to governmental liability in tort suits. Two major exemptions are: (1) "deliberate torts" (Section 421.6), and (2) the "discretionary function" exemption (Section 421.b). Since 421.b is a central legal issue in the atomic testing litigation, its limits must be clearly stated. The act's protections shall not apply to:

> any claim based on an act or omission of an employee of the Government, exercising due care, in the execution of a statute or regulation, whether or not such statute or regulation be valid, or based on the exercise or performance or the failure to exercise or perform a *discretionary function* or duty on the part of a federal agency or any employee of the government, whether or not the discretion involved be abused.

The Federal Tort Claims Act's "Discretionary Function" Exception

"Even if federal actors have patently abused their discretion, their performance of a discretionary function within the scope of their assignment moves them beyond the pale of the FTCA," stated the U.S. Supreme Court about this exception that the government has used in its defense in the recent atomic testing litigation in federal court.[25] In successful tort actions against the government, plaintiffs *must* refute the government's argument that the challenged actions are discretionary ones and hence exempt from the FTCA.

The question of discretionary function is the initial, major jurisdictional obstacle that the nuclear testing plaintiffs must overcome in order to make their substantive case for equitable remedy before the federal judge. The primary purpose of the "discretionary function" exception is to maintain the constitutional separation of powers and to prevent the judiciary from evaluating or interfering with discretionary actions of the U.S. Government's agents, especially in the areas of national defense and security. The exception, if upheld by the federal courts in tort actions, results in a governmental immunity from "liability for damages resulting from nuclear testing and research for the benefit of the public."[26]

The downwinders can overcome the government's defense only by showing that it was a *ministerial*, that is, a nondiscretionary activity, that caused the injury or death, and that the Atomic Energy Commission ignored standards of "ordinary care" that the FTCA requires of the government. While enabling routinely injured citizens to sue the government under certain circumstances, at the same time Congress "wanted to retain immunity for decisions made in the planning and administration of goverment activities." Therefore, the legislators "designed the discretionary function exemption to separate these two types of activities."[27] For the downwinders, there was an additional *practical* dilemma: neither the FTCA nor the federal court opinions have developed a "comprehensive definition of a discretionary function."[28]

Since 1953, the U.S. District Court judges, the U.S. Circuit Courts of Appeals judges, and the justices of the U.S. Supreme Court have confronted cases that enabled them to more precisely define the "discretionary function" exception in the FTCA.

In 1953, the U.S. Supreme Court set the tone for subsequent federal decisions in the lower courts when it interpreted discretionary function in a manner that gave the government broad immunity from tort suits. In *Dalehite v. United States*, 346 US 15 (1953), Justice Reed for the majority ruled the federal government not liable for an accident in a Texas shipyard that killed hundreds of people (in which the plaintiffs were asking for over $200 million in claims) even though the government agents were negligent. Introducing a distinction between *planning level* (nonliability based on the proper exercise of governmental discretion) and *operational level* activities of governmental agents, the Court majority stated that "where there is room for policy judgment and decision there is discretion. It necessarily follows that acts of subordinates in carrying out government operations in accordance with official directions cannot be actionable."

Reed concluded by stating that it "was not contemplated that the government should be subject to liability arising from acts of a governmental nature or function." Justice Jackson, dissenting, argued that the majority had weakened the FTCA to the point where it could only be used in minor tortious accidents involving motor vehicles, etc.; the FTCA, he argued, "embraces more than traffic accidents." *Dalehite* was a 4–3 opinion with two justices, William O. Douglas and Tom C. Clark, not participating.[29]

In *Indian Towing Co. v. United States*, 350 US 61 (1955), the U.S. Supreme Court itself modified *Dalehite* when it stated that once the government exercised discretion (to maintain a lighthouse) "and engendered reliances on the guidance afforded by the light, it was obligated to use due care to make certain that the light was kept in working order."[30]

Federal judicial opinions, including Supreme Court decisions, have developed the following guidelines in tort litigation involving the question of government immunity due to the "discretionary function" exception.

1. There is a case-by-case balancing of the conflicting individual and governmental needs.
2. Operational, mandatory duties or actions that prove to have been taken without due care are not immune from tort actions in federal court.
3. Once a policy is made and announced, over the entire period of its existence as public policy "due care" must be taken by operatives.
4. The government cannot (*over time*) "ignore the minimum requirements of ordinary care in the performance of a discretionary function."
5. Federal judges must "look beyond the words to the context in which a particular person may have been hurt and . . . determine if in that context the cost of that hurt to that person should be borne alone by that person or shared by all for whose benefit the hurt may have been inflicted."

There will always be these contextual, open-ended questions that the federal judges must answer. The federal courts have distinguished, in general "policy" terms, between policy-making and operational activities. But there is no definitive, comprehensive judicial statement defining "discretionary function" because the federal judges have to interpret the principles as they may apply in each suit *de novo*.

In these difficult cases, the federal judge, "must weigh administrative, economic, prophylactic and ethical factors."[31] Ultimately, the breadth of the judicial decree that follows such deliberation is related "to the capacity of the judge to engage in an effective analysis" of the conflicting policies that underlie official immunity from tort suits.[32]

The Federal Tort Claims Act and the "Causation" Question

Successful litigation of the negligence and wrongful death suits instituted by downwinders who contracted cancers and leukemias pivoted not only on the court's willingness to deny the federal government's immunity argument but on a requirement that is more difficult to establish. Rex Lee, then Assistant Attorney General, U.S. Department of Justice, conceded before a 1981 congressional committee that

> decades ago, federal policymakers decided to run some enormous risks. Innocent American citizens were involuntarily and unwittingly made the subjects of those risks and had thrust upon them the brunt of those risks.[33]

Assuming that this poignant request for equity were to be granted by a federal district court judge *and* that the judge declined to grant the government the immunity that accompanies the discretionary function exception, that federal jurist and the plaintiff's attorneys would still be confronted with the dilemma of *proving causation* in these nuclear-test tort suits.

Of the five conditions that must be established for the downwinders to win their civil suits, the most crucial is the *causation* of the plaintiff's illness by the radiation exposure."[34] The burden is on the plaintiff to show "more probably than not" that the cancer or leukemia was caused by the government's above-ground atomic testing program. In the downwinders' litigation, for example, the judge had to find that "the chances that the defendant's force caused the plaintiff's injury are at least slightly better than 50 percent."[35]

But, in this controversial tort liability suit, the harmed downwind plaintiff can only show an indirect causal linkage between exposure to low levels of radiation and the cancer that developed. In such a situation, wrote Stephen Soble,

> producing the evidentiary showing required to sustain the substantive proof of legal causation is an undertaking of no small magnitude. Logically, to prove causation, the plaintiff must be able to (1) isolate the harm-causing substance, (2) trace its pathway of dispersal from the polluter to the victim, and (3) show the etiology of the harm-causing substance. Without extensive scientific data these elements of causation cannot be firmly established. But introducing scientific studies—especially a full-scale epidemiological study—does not guarantee success in proving causation.[36]

Furthermore, in this radiation exposure tort suit, based on causal linkages, the (defendant) government lawyers will assuredly counter with its own government medical doctors, epidemiologists, and government scientists, who will certainly disagree with the plaintiff's scientific experts over interpretations of scientific data.

Another problem facing the downwind plaintiffs is the general reluctance of trial judges to dramatically expand the boundaries of tort law. There are many occasions where, in the "endeavor to address issues which often lie at the interface of scientific truth and reasonableness under tort law, a court may retreat from wrestling with scientific probabilities and uncertainties. . . . This [judicial] refusal to consider scientific issues is especially likely when the experts disagree."[37] For the plaintiffs' attorneys, these issues were troublesome ones that had to be dealt with head on in order for the plaintiffs to win their tort suit.

In these nontraditional tort cases the federal district court judge must decide whether to *accept as sufficient proof* the epidemiological studies or to reject them because of counter-testimony of the defendant's medical experts that challenges some of the statistics. Recall the admonition of Dr. Chase Peterson:"Scientific proof will never be 100%, and you should not be looking for it."

Whether a judge, who works with statements about evidence that are rather different from those of scientific research and scientific proofs, can accept Dr. Peterson's definition of scientific proof is critical to the determination of the atomic-test litigation. The dilemma of "limited scientific understanding of the etiology of cancer makes it almost impossible to establish proximate [legal] cause," wrote one legal observer. "Courts," she wrote, "have been reluctant to adapt tort requirements for proof of causation to accommodate scientific evidence."[38]

The lengthy latency period of cancer "makes it difficult to ascertain whether cancer is in fact the result of radiation exposure."[39] Because the Atomic Energy Commission failed to provide health hazard warnings, did not monitor fallout in the "food chain," and kept inaccurate dose records, the downwind plaintiffs must rely on statistical studies made years later (and subject to criticism) as the basis of proof that radiation exposure in the 1950s caused the onset of the cancer and leukemia in the 1960s and 1970s. If the federal judge rejects these studies, the plaintiffs are left without the *legal, proximate causality* required by the law of torts.

In addition to this critical judgmental dilemma, the federal judge must also deal with the statute of limitations issue. Ordinarily, a plaintiff must commence litigation under the Federal Tort Claims Act within two years of the date of the cause of action (28 *U.S.C.* Section 2401.b). Since the plaintiffs were unaware of the cause of their injuries, illnesses, and family members' deaths until documentation was made available to them in 1978 and 1979, did the statute of limitations clock start when the atomic clouds drifted over the downwind communities, thereby effectively barring these plaintiffs from seeking damages? Or does it begin when the plaintiffs discover that the tests caused the injuries and deaths?[40]

Yet another problem is the legal status of the unconceived plaintiff who suffered preconception injuries, "conceivably the most difficult and disturbing of the possible injuries."[41] These are injuries which a not-yet-conceived plaintiff receives as a result of genetic damage to one or both parents. The ramifications for tort actions are mindboggling! How does a federal judge, hearing an FTCA action and determining proximate cause, decide a case "in which the plaintiff was not in any sense in existence at the time of the tortious conduct?"[42]

Discretionary function, causality, latent injury, statute of limitations, and preconception injury are some of the major issues of the atomic fallout litigation. "Actions from injuries from radiation exposure," wrote a legal observer, "confront our legal system with issues as complex as the technology that engendered the widespread use of atomic energy."[43] Clark Downs, an attorney who testified before one of the congressional committees that examined the atomic testing issue, stated candidly that "the tort process is inadequate for this task."[44]

If rights have been violated (a critical and a hard judicial judgment) and the government is at fault (another judicial judgment), then "judges do not have the option of declaring that litigants have rights without remedies. The judge," wrote Federal Judge Frank M. Johnson, Jr., "has no alternative but to take a more active role in formulating the appropriate relief."[45]

Final Legal Strategy Sessions

By the end of 1978, with a great deal of information now available about the testing at the Nevada Test Site, the downwinders organized themselves into the Committee of Survivors. In late October, 1978, James Udall, son of one of the lead attorneys in the downwinders' effort to find an administrative or legal remedy for their government-caused injuries claim, came to St. George, Utah to help organize the group. *The Deseret News* reported, "Udall will spend the next several months working with the survivors of these cancer patients who may have been afflicted by disease because of the fallout problem."[46] The Committee systematically sifted through the thousands of complaints about injury and death allegedly due to the fallout radiation, and organized a small test group of plaintiffs who would bring suit against the federal government.

By August, 1979, after the Department of Energy had rejected their claims, the filing of the damage claims against the government (for leukemias, cancers, and other adverse health conditions) which the downwinders believed were caused by the radiation from the test, was accomplished. The federal judge selected to hear the *Allen* case was Bruce S. Jenkins, a local Utah citizen who had been appointed to the federal district court in Utah by President Jimmy Carter in May, 1978.

The Federal Judge: Bruce S. Jenkins

When the plaintiffs filed their papers in federal district court in Salt Lake City, Bruce S. Jenkins, fifty two years old, had been on the federal bench for less than a year. While he would hear some very controversial cases over the next few years,[47] the nuclear testing litigation would prove to be the most important, the most complex, and the most lengthy piece of litigation to come before him.

Jenkins grew up in Salt Lake City; his father was a local teacher and his mother, a court reporter. He attended the University of Utah, graduating Phi Beta Kappa with a degree in political science. Briefly undecided about a choice of careers—whether to pursue a Ph.D. in political science or a law degree—Jenkins finally decided on law and received his J.D. from the University of Utah in 1952.

Jenkins was always active in Democratic party politics,and after graduating from law school and establishing a private practice, he sought elected office. He ran unsuccessfully for Congress and for Mayor of Salt Lake City, and later entered the State Senate. In 1965, at age 37, he became the Democratic President of the Senate. That same year, Jenkins was appointed Federal Bankruptcy Judge for the U.S. District Court in Utah by the man he would later replace on the Court, Judge Willis W. Ritter. For thirteen years, he served as the state's federal bankruptcy judge, hearing over 17,000 cases.[48]

In March, 1978, Judge Ritter died, and President Carter, very committed to a merit selection process for federal judges,[49] created an eleven-person, Utah Federal Selection Commission. Over the summer they reviewed candidates and reduced the list of nominees for the federal position to five names: Jenkins; State District court Judges James A. Sawaya and VeNoy Christofferson; and Salt Lake attorneys J. Thomas Green and, interestingly, Dan Bushnell. (Bushnell is the lead attorney for the plaintiffs in the sheep death case, *Bulloch v. United States*.)[50]

Jenkins evidently received the nomination because he was recommended and supported by a number of leading state Democrats, including the only Democratic member of the congressional delegation, Congressman Gunn McKay, as well as the state's governor, Scott Matheson. On August 28, 1978, the Carter White House announced that the administration would send the name of Bruce S. Jenkins to the Senate for its approval. Both Republican Senators, Jake Garn and Orrin Hatch, complimented the Democratic President on the "wise choice" and indicated that they would not oppose Jenkins' nomination at the Senate hearings.[51]

In a candid interview with a *Deseret News* reporter on the day of the announcement, Judge Jenkins gave readers some keen insights into his judicial mind. Claiming that the courts were "very human institutions," Jenkins encouraged lawyers to do the best they could when they prepared a case. "You admire a case that's well-tried, a lawyer who prepares a case well (and who educates a judge well). *The law*," he continued, "*has great opportunity for creativity. The parallels aren't always found in the lawbooks*" [my italics]. Concluding, Jenkins stated that "implicit in any legal case are values. They're often unexpressed—but they're there."[52] The quality of justice, he said a few months later to a group of attorneys, "is just as much your responsibility as ours."[53] At another meeting of Utah attorneys, Judge Jenkins stated that "we all leave moral fingerprints on everything we touch."[54]

In an essay that appeared in the *American Bankruptcy Law Journal*, in 1974, on the nature of the judicial role, Jenkins wrote:

the judge is a steward of power. He holds power from the people in trust for the use of people. . . . Unlike any other type of public official, the judge, when confronted with an existing and concrete problem, is blessed with *ex post facto* wisdom and is twice blessed with existing power to deal with the specific problem . . . The job of the judge is to decide . . . judicial decision is choice. Such choice is not always between what is wrong and what is right, but is often between what is right and what is right—a difference in degree rather than a difference in kind.[55]

However, before making a choice, the judge needs to be *educated* by the lawyers before the bench. "At his finest, a courtroom lawyer is an educator," believes Judge Jenkins.[56] "The function of the courtroom lawyer is to convince the judge to use or to refuse to use judicial power. The mission of the courtroom lawyer is to have the power of the people, personified by the judge, stand with his client."[57]

In an intellectual showing and telling, the lawyer identifies the problem and then states what he wants the judge to do about the problem. "Next, as a lawyer, he needs to tell the court 'why' it ought to be done. He needs reasons for what he is asking."[58] At this point, the lawyer-educator walks the judge through the legislative and judicial judgments of the past that the lawyer has looked at carefully. He has already "made interpretive judgments long before he calls upon the court to do so. Only then is he in a position to provide guidance to the court; to point the way for the court. He can then demonstrate to the court not only how to solve the problem—but show what is in his mind the *best way* within the social framework for court power to be used in solving or resolving the specific problem."[59]

Finally, in addition to the *stare decisis* "whys" for judicial action, the lawyer-educator finds the answer in the "facts," and in "other reservoirs of knowledge beyond the customary experience. In philosophy. In religion. In common sense. In ethics. In economics . . . He finds answers in the recesses of his own mind and heart and in the processes of deliberate and purposeful thought."[60] And in the emotions. "Do judges have emotions?" asked Jenkins rhetorically. "Sure they do. The Court is a very human institution."[61]

Stewart Udall, chief counsel for the downwind plaintiffs, commented recently that Judge Jenkins "is a scholar and fairminded. I'm sure he has been aware that we presented a fair case."[62] One of his *Allen* responses reveals the jurisprudence of Judge Jenkins. Henry Gill, the chief counsel for the defendant, the U.S. Government, had argued on three occasions during the trial that there was no governmental negligence—"not even the garden variety negligence."[63] Three times Judge Jenkins rejected the government's arguments:

The Court's job is to look beyond the words to the context in which a particular person may have been hurt and to determine if in that context the cost of that hurt to that person should be borne alone by that person or shared by all for whose benefit the hurt may have been inflicted. We must focus to see if, at that point, in that distinctive context of time and place and people, government actions are of such a nature that persons who are hurt through government actions at that level should in good conscience be made whole.[64]

The Federal Judge's Dilemma in the Nuclear Testing Litigation

In the nuclear testing litigation, direct causality, i.e., that continual exposure to low levels of radiation produced by the atomic bombs led to the cancer and leukemia deaths, is almost impossible to show evidentially. Therefore, even if Stewart Udall showed governmental negligence and convinced the federal judge that administrative action was not exempt by the discretionary function exception in the FTCA, he still cannot receive a remedy in federal court if he cannot convince the judge that the fallout caused the cancer and leukemia.

If a federal judge, in such a tort suit, does go beyond the "threshold" jurisdictional question, i e , discretionary function, then the jurist is faced with deciding in favor of the plaintiffs, and thereby refashioning tort law from the trial court level or deciding against the plaintiffs, and ensuring that the thousands of plaintiffs in the suit would be without a legal remedy.

In tort-negligence cases, and the nuclear testing litigation is, at bottom, just such a litigation, the judge has to roam far and wide over the evidence to determine the justice of the matter. "Law," wrote Leon Green, "is the power of passing judgment." The judge, in making the law, has to go behind "the curtains of legal expression" and "enter the laboratories of his intelligence." These laboratories are not legal. "They comprise all we are . . . There we do our work. Out of all this comes a judgment."[65] Since the power of passing judgment is controlled by some factors that lay outside the books of law, "the hurt plaintiff [who] captures the heart of the judge" can be in a position to receive a judicially imposed remedy.[66]

In the downwinders' tort suit, U.S. district court Judge Jenkins was given the opportunity to pass judgment, to make law favoring plaintiffs, and to redesign the tort law with respect to discretionary function and causation. For Judge Jenkins, the questions of importance revolved around his perception of discretionary function and causality. Ruling for the plaintiffs on these issues, there are remedies available to heal the hurt.[67]

Judge Roger Foley in another 1982, Nevada Test Site case ruled against the plaintiffs. He chose not to push the law of torts in new directions. It certainly was not an easy task but, inevitably, it is a value judgment that the judges in our federal system have to make on occasion. As another federal district court judge wrote:

> Federal judges do not relish making hard decisions and certainly do not encourage litigation on social or political problems. But the federal judiciary in this country has the paramount and continuing duty to uphold the law. The courts must decide the issues—even though these decisions result in criticism.[68]

The plaintiffs in the *Roberts* case claimed a deprivation of rights but were unable to receive a remedy from the federal court. Judge Johnson noted: "Judges [if they conclude that rights have been violated in a case before them] do not have the option of declaring that litigants have rights without remedies. The judge has no alternative but to take a more active role in formulating the appropriate relief."[69] Judge Jenkins noted that the question is never "will the government win or lose? The government always wins when justice is done regardless of whether the verdict be guilty or not guilty."[70]

In our judicial system, decisions of great moment come down to the judgments of federal judges such as Bruce Jenkins. "An opinion," wrote Leon Green, "is but the smoke which indicates the grade of mental explosion employed" by the federal judge.[71] The downwinders' litigation, which reveals Jenkins' perception of the judicial role, also raises some serious questions about the capability of the judicial system, given the existing state of tort law and the limitations of the Federal Tort Claims Act, to respond to real injuries suffered by the downwinders who claim that the federal government's nuclear testing activities were the direct cause.

Scott M. Matheson, Governor, Utah. Governor Matheson has actively participated in the national effort to provide compensation for persons who have received injuries from the atomic testing fallout of the 1950s and early 1960s.

Dr. Joseph L. Lyon, Professor, Department of Family and Community Medicine, University of Utah. Dr. Lyon, a medical researcher at the University of Utah's Cancer Registry, has done extensive medical epidemiological work on childhood leukemias in southern Utah. He was an expert witness in the recent *Allen* litigation heard in U.S. District Court, Utah.

Chief Judge Bruce S. Jenkins, U.S. District Court, Utah. Judge Jenkins began hearing the *Allen* case barely a year after he became a federal judge in 1978. His opinion in the *Allen* case extended the scope of the Federal Tort Claims Act and is certain to be appealed to higher federal courts by the government.

Henry S. Gill, U.S. Department of Energy Special Counsel. Attorney Henry Gill has been the chief counsel for the U.S. Government in the *Allen* case from its first filing in 1979. He was disliked by the plaintiffs because of his abrasive courtroom behavior.

Stuart L. Udall, Attorney, Phoenix, Arizona. Attorney for the plaintiffs in *Allen v. U.S.*

Preston Truman,
Director, the Downwinders.

Janet Gordon, Director, Citizens Call, and her sister Mary Lou Melling. Janet has been the Director of Citizens Call since its inception
in 1978 and has been a rallying force for the people in the towns
downwind of the Nevada Test Site. Her sister has also been an out-
spoken advocate on behalf of the downwinders.

Fallout from an atomic test sweeps down on a working farm downwind of the Nevada Test Site.

"Dirty Harry" shot, Spring 1953. This shot was of one of the very bad atomic bursts, in that it deposited large doses of radiation, even by Atomic Energy Commission reporting standards of the day, on towns in Utah and Nevada such as St. George and Bunkerville.

On facing page:

Main Street, Cedar City, Utah, 1985 (top).

Kanab Middle School, Kanab, Utah, 1985 (middle).

U.S. Highway 89, Orderville, Utah, 1985 (bottom).

The Baneberry venting, Nevada Test Site, December, 1970. One of over forty known ventings of underground tests. This venting sent a radioactive cloud 10,000 feet into the skies above the Nevada Test Site. Over eighty workers were exposed to heavy doses of radiation. Three died of leukemia within a few years of the venting. The litigation is still developing in the federal district court in Nevada.

7

"We Are Going to Give Them a Fight": The *Allen* Case in Federal District Court

Every legal right should find a vindication in an effective remedy.

Norman Antin

The hazards of radiation are not reflected in case law or the statutes of the country.

Gordon Dean, Chairman, AEC

[The AEC] did not inform the public of the nature and extent of any hazards and of precautions which may be taken. They . . . neglected an important basic idea: *there is nothing wrong with telling the American people the truth.*

Bruce S. Jenkins

The Downwind Plaintiffs Go To Federal Court

Although Gordon Dean, an early chairman of the AEC, proudly stated at the beginning of the above-ground nuclear testing program at the Nevada Test Site that the "AEC has discharged its responsibility in such a way as to provide few opportunities for the tort lawyer,"[1] Stewart Udall and his co-counsel were committed to refuting this contention in the public forum. Udall, however, who had been around Washington, D.C. for many years as Secretary of the Interior in the Kennedy and Johnson administrations, "recognized reality."[2] The reality for the former cabinet official was the existence of the the 'old boy' network in the bureaucracy and Office of Management and Budget. In testimony before Congress, Udall stated, "The statements [of the Reagan Administration about the fallout litigation and compensation problem] are different only in a small degree from what the Carter administration presented 18 months ago. Do not let them cross the threshold . . . They are protecting the nuclear establishment."[3] The plaintiffs' lead attorney was cognizant of the strong and influential policy triangles on Capital Hill, i.e., the policy network that exists between the lobbyists, the federal agency bureaucrats involved in nuclear testing, and the legislative staffers on the Hill, but he was committed to the task of making a strong and decisive case. "We are going to give them a fight," he asserted before a congressional committee.[4]

145

On August 30, 1979, the *Irene Allen v. United States* suit was filed in the U.S. District Court by 24 plaintiffs. These plaintiffs were the first group of almost 1,200 people seeking remedial judicial action for injury and wrongful death from exposure to radioactive fallout resulting from the negligence of the (defendant) United States' atmospheric atomic testing program from 1951 to 1962, under provisions of the Federal Tort Claims Act.[5] The plaintiffs, residents of Utah, Nevada and Arizona, were the surviving spouses and children; representatives of deceased individuals; and present "victims of leukemia, cancer or other radiation-caused diseases or illnesses."

The plaintiffs charged the federal government and its agents with negligence in the preparation, testing, evaluation, control and conduct of the atomic tests and/or the failure to act with due care, resulting in the injury, disease, and deaths of residents of the downwind area. These injuries and deaths, they claimed, were "caused by exposure to radiation resulting from defendant's atmospheric . . . atomic/nuclear testing program during the time indicated."

Specifically, the downwind plaintiffs maintained that the federal government was obligated to use "reasonable care" to warn citizens about the hazards of radiological exposure and to thereby allow them "to fairly evaluate the risk to them of atomic experimentation." The government was "under a continuing duty to conduct each test in a reasonably safe manner so as not to endanger human life or property which might be exposed to and harmed by radioactivity created by the atmospheric detonation."

The case was predicated on the fact that the government "failed to use reasonable care" in the activities associated with the nuclear detonations at the Nevada Test Site, and charged the government with negligence in fulfilling its "active and ongoing duties." As a result of negligent acts and omissions by various government agents, "plaintiffs deceased, and plaintiffs were exposed to a sufficient amount of harmful radiation to cause death or plaintiff's present illness."

Complaints for administrative settlement with the Department of Energy (filed in the fall of 1978) had been denied because they were pending before the department for more than six months, so the plaintiffs had to seek a remedy in the federal district court under the provisions of the Federal Tort Claims Act.

The government's response, filed October 30, 1979, with the U.S. District Court, urged Judge Jenkins to dismiss the suit because the plaintiffs' complaint failed to "state a claim upon which relief can be granted." The government's attorneys, led by Henry Gill, urged the federal court to dismiss the case on the following grounds: (1) the discretionary function and misrepresentation exceptions immunize the government, and therefore there is no federal court threshold jurisdiction in the matter; (2) there was no negligence in the government's implementation of the above-ground atomic weapons testing program at the Nevada Test Site; and (3) there was no direct legal causality shown, i.e., "there is no competent scientific evidence that any exposure to radiation resulting from defendant's test activities caused the illness or disease alleged by petitioners."

Discovery in the *Allen* Case

The Federal Rules of Civil Procedure, devised by Congress in 1938 to speed up the civil litigation process in federal courts, included the discovery process, which allows both parties, under court supervision, to engage in a total review of the facts and to examine the witnesses in order to find out what both sides will confront in the litigation effort. It is a basic pre-trial strategy that encourages the parties to think settlement rather than to proceed with the litigation to the formal trial stage of the civil action.

The extremely complex and time-consuming discovery process of the *Allen* case continued from fall, 1979 through spring of 1982. Both sides interviewed dozens of witnesses. The plaintiffs' efforts focused on the Atomic Energy Commission's scientific knowledge of the hazards of radiation exposure. The defendant focused on the plaintiffs' state of health, the general scientific question of causality, and the question of whether the federal government had the discretionary immunity to conduct above ground atomic tests in the 1951 to 1962 period.[6]

The going was tough for the plaintiffs' attorneys during the discovery phase of the trial. Before going to formal trial, both sides explored the possibility of settlement,[7] and Udall said he "personally favored a negotiated settlement."[8] But this was not to be the case in *Allen*. The defense attorneys for the government were taking an "extremely hard line posture" that reflected, for Udall, "the heavy hand of the Office of Management and Budget." "'Don't let them in the door,' they counsel, 'because we don't know where it will lead.'"[9] The Carter Administration's refusal to even discuss settlement of the nuclear fallout cases illustrated Dr. Lyon's belief that governmental judgments about atmospheric testing, cancer, causation, and compensation are totally inseparable.

Udall recalled visiting with people in the Carter White House about the litigation. "Briefly, naively, I thought maybe we could talk settlement, settlement at least of the children that Dr. Lyon had identified in the leukemia epidemic, that maybe the government would step forward and we could begin to heal this whole process."[10] This was never to be, either with the Carter people in the Office of Business Management or with the Reagan administration. As Udall somberly stated recently: "I am past anger in all this . . . They are saying, 'stick it to them. Make them prove it. Go to court.'"[11] When discovery ended, Udall and his staff, and the attorneys for the government, went about the task of developing their trial strategies.

The Role of the Lawyers in the Search for Justice in *Allen*

It is clear from a review of the interrogatories, depositions, and case strategy notes prepared by the plaintiffs' attorneys, that the plaintiffs' attorneys in *Allen* understood the conception of justice discussed by Judge Jenkins in his law review essay published years earlier.

Between 1979 and 1981, Udall received memos from two of his staff attorneys, Chris Cannon and Craig Carlisle, which reveal that they possessed a practical understanding of Jenkins' admonition that lawyers work with the judge in reaching justice. The memos also show that careful attention was being given to both the Ninth and Tenth federal circuits' past judicial judgments regarding the threshold issue of discretionary function. Beyond the review of precedents, the memos candidly discussed the difficulty the attorneys would have in breaking new ground with the concept of discretionary function. "There is an egregiously pro-discretionary function exemption inclination in the tenth circuit," Cannon cautioned Udall in July, 1979, although the ninth circuit "decisions are more rational."[12] Their research effort, they privately admitted to each other, had to "draw the golden thread which will avoid the pitfalls of *Dalehite* and other cases where a discretionary function was found."[13]

The final strategy took shape in these early conversations: show and tell the judge how vague the operational plans were because "the more vague the plan, the more important the discretionary aspect of each decision."[14] They reasoned that "if a plan includes a direction to care for the public health without more, the discretionary nature of that element does not protect the lower level decisions.[15]

"Operational mismanagement" was not protected by the discretionary function exemption, they reasoned. Therefore, their premise was a simple one (a premise accepted by Judge Jenkins in his opinion in May, 1984):

> It is not that atomic weapons should have been tested, but that the tests were so negligently and wrongfully carried out that the government should be liable for the harm to the health of people in the affected areas[16] . . . Throughout the argument we must emphasize the lack of policy considerations and attribute the decisions not to warn to lower level personnel.[17]

The plaintiffs carefully developed an educational program for the judge. The government's position, however, was much more diffuse, and its case was not as cohesive as the plaintiffs' arguments before the federal judge (perhaps because there were two groups of federal attorneys—from Washington, D.C. and Utah—and they were not in contact as regularly as the plaintiffs' attorneys).

In an interview in late January, 1984, Henry Gill, a hard-nosed legal expert for the Department of Energy, claimed that the government's strategy in this case was to argue an absolute immunity from tort liability—down to immunity for the "garden variety" negligent acts of government officials. His effort to educate the judge was "structured by the testimonial evidence,"[18] not by a serious effort to strengthen the judge's comprehension of discretionary function.

Armed with clear instructions from both the Carter and Reagan Justice Departments (and the Department of Energy) to stand firm on the issues of discretionary function and causality, a stone-walling, reactive group of defense attorneys toiled before Judge Jenkins. Instead of leading the judge through the pathways of the law and facts on the issue, they reacted inflexibly to his questions from the bench.

Poor planning undermined the government's defense. For example, Gill himself indicated his *reactive* strategy in a number of incidents. When a government witness was being questioned by Jenkins about cancer and children, Gill was not prepared to rebut and had to delay in order to gather evidence to buttress the government's case in court—but not until the following day.[19] Another example of the government's poor planning is the assertion that while Dr. Charles Land, of the Center for Disease Control in Atlanta, was dining in New Orleans with some of his staff, he was given one week in which to prepare rebuttal testimony for the trial to refute Dr. Lyon's testimony.

In its responses to plaintiffs' interrogatories there were indications that the government used "propaganda" rather than truth in educating the judge and the attorneys. For example, in the government's response to the October, 1979 questions, the attorneys wrote that there was "no competent scientific evidence that any exposure to radiation resulting from defendant's test activities caused the illness or disease alleged by plaintiffs."[20]

In November, 1979, the government made the following response to the question of whether the government had the responsibility of notifying its citizens of potential hazards: "If responsibility is a description of a legal duty, the United States denies it had the responsibility of notifying its citizens of any potential health hazards. That determination [not to warn the citizens] is a matter properly the subject of executive discretion."[21]

The records show that the government's strategy was not to move away from existing precedents regarding discretionary function and proximate causality. From the beginning of the litigation, it was obvious that the government attorneys, carrying out an inflexible policy from the White House, were no match for the plaintiffs' counsel in judicial creativity.

Allen v. United States: The Defense Motions to Dismiss

In April, 1981, the government lawyers filed a motion in the federal court to dismiss the case for lack of jurisdiction, claiming the government agents at the Nevada Test Site in the 1950s were immune from tort suits due to the "discretionary function" exception in the Federal Tort Claims Act. This was to be the first of three such motions introduced by the federal government's attorneys to dismiss the suit for lack of jurisdiction. In August, 1981, these attorneys filed in court a renewed motion for dismissal. In it, Henry Gill, the lead U.S. attorney, argued that the *entire* atomic testing program, from "high level policy decisions" to the "garden variety negligent acts or omissions, are immune from liability" under the discretionary function exception in the Federal Tort Claims Act. "*There are no operational level acts not within the scope of the discretionary function exception,*" the federal government lawyers asserted categorically. In addition to deriving governmental immunity under 28 *USC* Section 2680.a—discretionary function—the government argued that, even if there were negligence due to the lack of clear hazard warnings, the FTCA's misrepresentation exception, Section 2680.h was applicable and effectively barred claims against the federal government. Furthermore, the statute of limitations had expired. For these basic legal reasons, Gill maintained, there was no standing to sue on procedural grounds, and the government attorney requested Jenkins to dismiss the suit.

In August, 1981, in *Allen v. United States*, 527 F Supp 476, the federal judge denied the government's motion to dismiss. Judge Bruce Jenkins concluded that the record was not yet factually sufficient for the court to determine the issues raised in the filings:

> This Court cannot dismiss these cases wholesale at this stage of this proceeding upon the basis that each action up and down the ladder of governmental activity in conducting the testing program over a period of years was 'discretionary,' and thus insulated the government from answering in damages for alleged misconduct in carrying out the testing program. At this point we don't know enough. In a case of this magnitude *fundamental fairness* compels that any decision made be based on all available information. Thus, this Court now denies

the Government's motion to dismiss based upon the discretionary function exception without prejudice to its later reassertion upon a more complete record.

Subsequent to this ruling, discovery continued and plaintiffs continuously and unsuccessfully sought an out-of-court settlement with the government. A trial was set to begin on September 20, 1982. On September 13, 1982, both sides filed pre-trial statements with the federal court. The plaintiffs' document focused on the information the attorneys had uncovered during the discovery proceedings. This information seemed to indicate negligence, fraud, and deception committed by the Atomic Energy Commission over the eleven-year period of above-ground atomic testing in its failure to warn the plaintiffs of the danger of fallout.

The plaintiffs' brief indicated that there was a "preponderance of evidence that the injuries were caused by the negligent acts of the United States"; that the government had certain obligations and duties that were not followed; that there was a proximate causal connection between the governmental conduct and resulting injuries; and that there was actual loss and damage. Their *Findings of Fact and Conclusions of Law*, submitted in October, 1982, attempted to substantiate these points.

Using the *Indian Towing* Supreme Court decision (1955) as a precedent the plaintiffs argued, in their conclusions of law, that the injuries were the result of operational judgments not immunized by the discretionary function exception to the Federal Tort Claims Act: (1) Negligence occurred at the operational level and, according to established case law, once the duty is established, the planning and the execution of details were not discretionary; (2) "due care" was not practiced by the AEC during the above-ground testing period; and (3) the "preponderance of evidence" standard in civil cases meant that the plaintiffs' burden was to show that the radiation exposure was a cause, not that it was the sole cause of the cancers and leukemias. Since the preponderance of scientific evidence was that cancers were caused by radiation fallout, proximate causation was established.

In early September, Henry Gill, the government's attorney, asked the court again to dismiss the case on "discretionary function" grounds and, on September 13, 1982, was again rebuffed by Judge Jenkins who told the government that he wanted a "full-blown trial" in the downwind civilians' case. Whatever the decision, Judge Jenkins noted it was certain to be appealed to higher courts and it was important that a full record be present to understand the "complete" story behind the plaintiffs' allegations.

In their *Proposed Findings of Fact and Conclusions of Law*, filed in October, 1982, the Justice Department lawyers insisted again that the governmental actions were not negligent, that there was no proximate causation shown, and that, in any event, the government was immune from such tort suits because of the discretionary function exception.

Allen v. United States: The Trial

The trial began on September 20, 1982, and continued for almost three months until the summary arguments were concluded on December 17, 1982. The plaintiffs' attorneys presented dozens of witnesses, including medical researchers, the plaintiffs themselves, and former government employees, to show governmental negligence, to introduce epidemiological studies that showed proximate causality, and to reveal the types of non-immune local judgments made by the Atomic Energy Commission staff that led to injury and death.

Dr. Joseph Lyon, the medical epidemiologist at the University of Utah, was a major witness. His important testimony indicated that childhood leukemia deaths in Utah increased by 40% in the early 1960s, and that the counties in southern Utah, in the path of the atomic radiation clouds, had an increase in such deaths that was 2.4 times higher than the rest of the region, while the five counties closest to the test site had an increase that was 3.4 times higher than all other areas. Asked by the judge to discuss his findings in terms of probability, Lyon replied that radiation was the probable cause of the excess deaths, and contributed to a 71% probability that the young people in the counties closest to the test site had contracted cancer and leukemia as a result of the atomic-testing radiation.

Plaintiffs took the stand to recount their own tragedies and those of their loved ones who had already succumbed to cancers and leukemias caused, they strongly believed, by the radiation from the atomic tests that contaminated their lives.

Defense attorneys did not endear themselves to the plaintiffs during the trial. At the end of the first day of testimony by a plaintiff who was suffering from cancer and who had taken off her wig to show the ravages of her chemotherapy, one allegedly said something in the corridor outside the courtroom that made Janet Gordon, co-director of Citizen's Call, livid with rage. According to Gordon, he strode out of the courtroom and "in an arrogant voice," spoke of suing the government because he had lost some hair. [22]

Former employees of the Atomic Energy Commission and U.S. Public Health Service, such as Frank Butrico (one of several Public Health Service monitors who worked in St. George) testified about the lack of warning standards and, more ominously, that the Atomic Energy Commission had doctored reports which stated that the AEC had "excessive" standards for radiation exposure. He spoke of receiving no warnings from the AEC about detonations, and of never having been informed about "safe" radiation levels or what to do in case of a nuclear emergency.

Butrico also testified that a 1953 nuclear test report issued in his name was not prepared by him and contained inaccurate information. "Liberties" were taken with times and dose levels, he stated. In addition to placing him at improper locations during the May, 1953 "Harry" shot, the "Butrico Report" claimed that school children were inside when the radioactive cloud went over St. George—not true—and that Butrico's readings were within the AEC's "acceptable" radiation exposure range of 3.9 roentgens—also untrue. Butrico's actual fallout reading was 11.5 roentgens.

In mid-October, prior to developing the government's defense, the U.S. Justice Department attorney, Henry Gill, moved to dismiss the case for a third time because of lack of jurisdiction over subject matter, discretionary function, and no showing of causation. He forcefully maintained that there was "not one scintilla of evidence introduced in these five weeks of trial that even addressed the issue." No negligence had been shown, the attorney maintained.

In response, the plaintiffs' attorney, Ralph Hunsaker, argued that negligence had been shown if negligence means "doing something you shouldn't or failure to do something you should." The defendant did know about the radiation effects, and the AEC was negligent in their on- and off-site administration of the program. Causation was shown through the use of epidemiological studies.

Hunsaker also discussed the statute of limitations issue: "people there did not and could not have reasonably known about any potential claim that they may have had." Hunsaker distinguished the discretionary activity of the operations people at the Nevada Test Site and those assigned to nearby off-site locations (the monitors, information officers, off-site radiation officers, and test managers) from the high-level discretionary judgments of the President and the AEC Commissioners. The regional and local governmental agents were responsible for taking "due care" in the performance of their tasks and were not immune from liability, whereas, concluded Hunsaker, the high-level planning discretions of Truman, Eisenhower, and the AEC were immune under the "discretionary function" exception.

For the third time, Judge Jenkins ruled against the Government attorney and said:

> I think that there are enough terribly serious questions raised in this lawsuit, not just the legal issues but the related social issues. I think what I'm going to do here is to follow the power that the Rule [41] talks about and decline to enter

any judgment until the close of all the evidence . . . To do any kind of an adequate job we have to know not just a portion of the story but the whole story
. . . There is some virtue in allowing the United States to proceed to tell . . .
their version of what the consequences were . . . That way, a fact-finder then
can be in a position to be at least comfortable in making the kinds of factual
determinations that need to be made and drawing the kinds of conclusions that
ought to be drawn . . . Very, very serious questions have been raised.

The government's defense was based on a series of witnesses who testified
that the AEC's guidelines were not negligently drawn and that the plaintiffs'
cancers were not caused by the atomic testing program. Other experts on
radiation physics and dosimetry testified on the medical issues associated
with the fallout litigation. Judge Jenkins, after carefully listening to this type
of testimony for several days, commented from the bench: "I have had an
introductory course [in nuclear physics]; maybe they can give me an
advanced course."

Gill, the government's lead attorney, placed defense witnesses on the
stand who testified about the reasonableness of the Nevada Test Site's safety
standards, the care evidenced by the governor's off-site monitors as they
performed warning tasks regarding the off-site radiation exposure, the weak-
nesses of the plaintiffs' epidemiological studies, and how government officials "carried out that testing in conformance with the delegations that they
received from the President."

On December 17, 1982, both sides summed up their cases before Judge
Jenkins. The plaintiffs' attorneys focused again on the "grave wrongs" committed against the downwind plaintiffs because the government operators at
the Nevada Test Site had acted in a grossly negligent fashion. Hunsaker concluded with a comment that underscored the social and ethical dilemma:
"We're talking about people—individuals who died; we're dealing with the
most horrifying of injuries—slow, painful, agonizing, almost always terminal
injury."

Gill and other government attorneys summed up the AEC's case by raising
the statute of limitations issue and the government's absolute immunity from
liability under the Federal Tort Claims Act, and by claiming that the plaintiffs
did not demonstrate the AEC's negligence.

Judge Jenkins took the matter under advisement. Seventeen months later,
after he had reviewed the 7,000-page trial transcript, 54,000 pages of
exhibits, five boxes of depositions, and had taken "judicial notice" of ele-
mentary principles of physics, chemistry, and other sciences, as well as government compilations of scientific and historic data, Judge Jenkins handed
down his order in the historic case of *Irene Allen et al. v. United States*: it was
a landmark judicial opinion.

Irene Allen et al. v. United States: The Landmark Jenkins Opinion

Judge Bruce Jenkins' *Allen* opinion set a precedent for this type of tort liability suit. He broke new ground in his assessment of discretionary function and proximate causation in litigation involving complex, scientific questions associated with liability for novel injuries and wrongful deaths. His opinion held that the government's operatives at the Nevada Test Site did act negligently and were more likely than not responsible for the deaths and injuries of ten of the twenty-four plaintiffs. It was the first ruling by a federal judge against the federal government in a tort liability suit involving individuals who had received injuries or died from exposure to low-level radiation produced by a U.S. agency.

Jenkins worked on the case for seventeen months with his law clerk, Russell Kearl (who remained with the judge for nine months beyond the end of his clerkship to provide Jenkins with legal expertise in this highly technical and controversial litigation), and sacrificed weekends and evenings to learn the complexities of nuclear physics, biochemistry, statistics, and medical epidemiology.

The judge and his law clerk labelled room 241 of Salt Lake City's U.S. Federal Courthouse the "Theoretical Physics Division, U.S. District Court." Jenkins and Kearl worked tirelessly there, developing what Jenkins termed a "rational framework" and a "reasonable" judgment based on the available data that cut across so many fields of knowledge.[23]

The result of their efforts was an extremely well-crafted, massive, 215-page document that addressed the major issues in the litigation: "This case," wrote Jenkins, at the beginning of his dissertation, "is concerned with atoms, with government, with people, with legal relationships, and with social values."[24]

The federal judge also stated that, "at the core of this case is a fundamental principle—a time-honored rule of *law*, an ethical rule, a moral tenet: 'The law imposes a duty on everyone to avoid acts in their nature dangerous to the lives of others.'"[25] All the specific rules of negligence and proximate cause that the judge used as the basis for liability in *Allen* "are rooted in this principle of duty."[26]

In *Allen*, "as in any other case in tort law, the answer to the ultimate question: 'Who should bear the burden of the risks created by the defendant's conduct?' is ultimately a question of policy and of public value."[27] There is a clear admonition to the legal and political community that risks were involved in the atomic testing program at the Nevada Test Site, and that if the plaintiffs were involuntarily and negligently placed at risk without fully understanding these risks and were injured as a consequence, then *someone* had to bear the burden for the costs of the injuries and wrongful deaths.

To answer the questions of duty, breach of duty, risk, and liability Jenkins started at the beginning. His first 50 pages provide for the reader a brief, well-documented introduction to the history of radiation, nuclear physics, nuclear fallout, and health physics. In order to understand the legal and political nature of the *Allen* litigation, it was imperative for him to wrestle with these complicated scientific and medical issues. Jenkins outlined the relationships between the nuclear blast, the ionizing radiation accompanying it, the theoretical assumptions about the impact of radiation on cells, especially on DNA, and the associations between low doses of radiation fallout and the onset of cancer. He then turned to the important legal questions of discretionary function, statute of limitations, duty and breach of duty, and critical questions of causation and damages.

The critical threshold jurisdictional question for Jenkins was the issue of the discretionary function exemption. The government attorneys, led by Gill, had urged the judge to rule that the federal court did not have jurisdiction because the actions taken by Nevada Test Site operatives were covered by the discretionary function immunity. Looking "beyond the words" of the Federal Tort Claims Act exemption, Jenkins examined "what the United States did or did not do, about which plaintiffs complain" to determine whether these actions "were so imbued with considerations of public policy and governance as to be immunized from the reach of the FTCA."[28]

The plaintiffs were not complaining about the decision to test atom bombs above-ground on the Nevada desert. Instead, they were arguing that the testing program was administered in a negligent way which led to wrongful injuries and death—subject to tort liability judgments against the government.

The government argues that once a policy choice is made, all subsequent operational choices are immune from tort action, and evidently "misperceives the intent of the act." This operational choice is "subject to a standard, a limitation. It is subject to a standard imposed by a civilized society as to *appropriate conduct.*"

Operational judgments at the Nevada Test Site are not matters of discretion that are immune from tort liability actions in federal court.

> The manner in which the tests were conducted, carefully or carelessly, was also a matter of choice, but was not a matter of discretion because such operational conduct was subject to a standard, a limitation. That limiting standard of conduct, due care, reasonable care under the circumstances, is called a duty. The person to whom a duty is owed is said to have a right. At the operational level employees of the United States had a duty to prepare and conduct tests carefully with full regard for public safety. The citizen adjacent to the testing site had a right to have that duty fulfilled.[29]

The downwind plaintiffs complained in federal court that the duty was not performed and therefore that their rights were deprived at the cost of injury and death. "They complain of what the defendant didn't do—namely, that it did not adequately warn, did not adequately and contemporaneously measure, and did not adequately educate the population at hazard in simple and inexpensive preventative and mitigating measures."[30]

Jenkins thus concluded that these operational actions, not immunized by the discretionary function exemption in the Federal Tort Claims Act, were "negligently insufficient."[31] Therefore, he concluded, the "discretionary function exception for the operational activities complained of is simply unavailable. Jurisdiction is proper."[32]

Jenkins then carefully but cleanly dispatched the government's claim that the statute of limitations had expired. Knowledge, he said, starts the two-year limitations clock running. When did the downwind plaintiffs develop the knowledge that there was a causal relationship between their cancers and leukemias and the radiation fallout? Since this poses a complex question for scientists, Jenkins concluded that "no genuine social purpose that I can see is served by the imposing of a *procedural* bar, the statute of limitations, when the facts of knowledge of injury and human source have been such a long time coming."[33]

Next, the judge raised the question, did the federal government have some kind of legal duty to the persons downwind? Jenkins concluded that an absolutely clear legal duty was embodied in the Atomic Energy Act of 1946. The Atomic Energy Act "makes repeated, express reference to the protection of health and safety as a significant goal for activities of the AEC created by that Act."[34] The Atomic Energy Act of 1954 "reinforces the concern for health and safety expressed in the 1946 legislation."[35] Jenkins also concluded that, together, these acts imposed a duty upon the "government—the AEC—in favor of the plaintiffs to avoid the specific risks of injury at issue in this lawsuit."[36]

The government established and applied a reasonable standard of care at all of their nuclear laboratories except the Nevada Test Site. If stringent care is employed at these places, stated Jenkins, "why should not at least as much care be exercised in dealing with the . . . radioactivity in the pink-orange clouds of dust, gases and ash drifting eastward from southern Nevada?"[37] Because of governmental negligence and failure to warn plaintiffs of the health risks associated with radiation, the downwinders were involuntarily placed at risk.

The government had "vastly superior knowledge" about the risks of radiation exposure and therefore a "stringent duty of care to minimize such hazards."[38] Therefore, Jenkins held "that in its conduct of open-air atomic testing, and in its off-site radiation safety programs—including its public information activities—the government was bound by a legal duty to act with the highest degree of care in light of the best available scientific knowledge."[39]

The question that naturally followed was whether or not the duty to act with stringent care was breached by the government operatives. Jenkins reviewed the various "technical documents and reports prepared by NTS personnel following each of the nine major series of atmospheric tests,"[40] and related material, such as radiation safety plans, public information plans, and documents related to the shots, and general reports, summaries, statements and articles discussing radiation fallout exposure,[41] all of which confirmed an "astounding fact":

> At no time during the period 1951 through 1962 did the off-site radiation safety program make any concerted effort to directly monitor and record internal contamination or dosage in off-site residents on a comprehensive person-specific basis.[42]

Jenkins concluded that if careful, person-specific monitoring had been done and records had been kept, "accurate estimation of actual dosage to individual persons could have been achieved"[43] and proper warnings could have been issued.

In sum, the downwind plaintiffs were never really told that "*radiation exposure should be kept as low as practically possible because any degree of exposure to ionizing radiation may involve some degree of risk.*"[44] The Atomic Energy Commission failed in its duty to alert the citizens to the dangers of radiation exposure: "Consequently, many people were exposed to more radiation, and greater risk, than ever needed to be,"[45] and were deprived of the opportunity to protect themselves.

Jenkins was convinced, after reviewing the documents, that

- the AEC programs at the NTS—public safety and monitoring fallout— were badly flawed;
- the information given to the downwinders was woefully deficient;
- there was a failure to inform;
- there was a failure to monitor adquately;
- there was a failure to explain the increased risks to children;
- there was a failure to warn about cattle feeding and food chain contamination;
- there was a failure to advise people who were at risk about simple things they could do to minimize danger;
- and there was general failure to inform the public of the nature and extent of the hazards, and of the precautions they could take to minimize the danger to themselves and to their children.[46]

The AEC operatives "negligently and wrongfully breached their legal duty of care to plaintiffs as off-site residents placed at risk."[47]

Given his conclusion that the Atomic Energy Commission breached their legal duty of care to the downwinders, Judge Jenkins turned to the critical, problematic, and substantive issue of causation: Did the AEC's negligence and breach of duty more likely than not cause the plaintiff's leukemias and cancers? In a tort action for damages, said Jenkins, "each plaintiff must show that he has suffered injury as a result of the defendant's conduct, at least in part" and must "demonstrate factually that there is a reason why this particular person is the defendant."[48]

In this radiation litigation, Jenkins admitted that the "factual connection is very much in genuine dispute."[49] In ordinary tort cases involving *direct causation*, he noted that the "cause is far more direct, i.e., A fires a gun at B, seriously wounding him," but in the *Allen* litigation which involved the problem of *indeterminate causation*, "A irradiates B, who develops a tumor 22 years later."[50]

Proving cause-in-fact in such a complex, indeterminate causation tort case is very difficult because of "the nature of the injuries suffered," the "nature of the causation mechanism," i.e., the (unobserved) impact on human tissues of radiated particles ingested by downwinders, and the "extraordinary time factors involved" in the tort litigation, i.e., the lengthy latency period of the disease.[51] Judge Jenkins pushed forward, saying that despite the difficulty of examining the causation arguments, the Court "must use its own best judgements, experience and common sense in light of all the circumstances. This is true even in cases when it may be extremely difficult to establish a factual connection."[52]

In the precedent-setting portion of his opinion, Jenkins said in a radiation case such as *Allen*, which involves the problem of indeterminate causation, "a remedial framework can certainly be fashioned to meet the circumstances and requirements of the parties and issues now before this court in this action."[53] Jenkins then held that:

> (1) When the defendent negligently creates a radiation hazard which puts an *identifiable population* at risk, and (2) a *member* of that *identifiable population* develops a biological condition which is consistent with having been caused by that hazard to which he has been negligently subjected, such consistency having been demonstrated by substantial, appropriate, persuasive and connecting factors such as: the probability that plaintiff was exposed to ionizing radiation due to fallout from the NTS in rates in excess of natural background radiation; plaintiff's injury is of a type consistent with those known to be caused by radiation exposure; plaintiff resided in the area downwind of the NTS between 1951 and 1962; and other factors, e.g., time and extent of fallout, age, radiation sensitivity

factors, etc., (3) the *fact-finder*, i.e., the federal district court judge, may reasonably conclude that the hazard caused the condition absent *proof to the contrary offered by the defendants"* [my italics].[54]

Jenkins stated that each plaintiff in *Allen* "must establish," by the preponderance of evidence standard, at least the first three of the above factors—exposure in excess of background radiation; injury that could have been caused by radiation; and residence in the area during the period of above-ground testing at the Nevada Test Site. If the plaintiff could present the kind of evidence that would enable "reasonable men to conclude that it is more probable that the event was caused by the defendant than that it was not," then the law could provide a remedy for damages and wrongful death.[55]

In this showing by the plaintiffs, Jenkins noted that mathematical proof "is not itself the answer to these questions."[56] Nor need it be proof beyond a reasonable doubt, noted the federal judge.[57] What the federal judge used, in evaluating the facts in light of his remedial framework, was the "more probable than not" standard.[58] What he found overturned the existing precedents in indeterminate causation tort litigation in federal courts.

Jenkins analyzed the many facts presented at the trial by each of the twenty-four "bellwether" plaintiffs and concluded that the governmental operatives at the Nevada Test Site were engaged in "risk taking conduct," and that some of the plaintiffs had received injuries "consistent with the harm that is predicted and observed when such risks are created. . . . consequently, the factual situation seems even more exclusive—exclusive of other defendant's other connections, other causes.'"[59]

Liability was imposed on the government in cases in which the plaintiffs had been exposed to excessive amounts of radiation from fallout, had cancers or leukemias of a type consistent with those known to be caused by ionizing radiation, and had actually lived downwind of the Nevada Test Site between 1951 and 1962.[60]

After carefully examining the medical records and other records, Jenkins determined that ten of the twenty-four plaintiffs suffered and nine had died from cancers and leukemias resulting from radiation exposure. "It appears," Jenkins concluded, "that ten of the twenty-four bellwether cases merit compensation. Eight are wrongful death cases . . . [and] the two additional cases which merit compensation are brought by persons living at the time the action was commenced who claim personal injury to themselves."[61] Jenkins reviewed the Utah and Arizona wrongful death case law (federal judges in Federal Tort Claims Act litigation apply the appropriate state law when they make the damage determination), considered the loss of assistance and support to the family, the surviving spouses' and children's loss of companionship and happiness, loss of the possibility of inheritance,[62] and announced the total award: $2.66 million to the survivors of the decedents in the Allen case:

Decedents	Awards to claimants	Type of Cancer
Arthur F. Bruhn	$625,000	Leukemia
John E. Crabtree	$234,000	Leukemia
Karlene Hafman	$250,000	Leukemia
Sybil D. Johnson	$250,000	Leukemia
Lenn McKinney	$300,000	Leukemia
Sheldon Nisson	$250,000	Leukemia
Peggy Orton	$250,000	Leukemia
LaVier Tait	$400,000	Leukemia
Living Claimants		
Jacqueline Sanders	$100,000	Thyroid
Norma J. Pollitt (now dead)	Amount not yet fixed	Breast

After announcing the damage awards, Judge Jenkins concluded his lengthy decision by summarizing the findings of fact and the conclusions of law. There was gross negligence by the federal operatives at the Nevada Test Site. "As a direct result of such negligent failures, individually and in combination, defendant unreasonably placed plaintiffs or their predecessors at risk of injury and, as a direct and proximate result of such failures . . . plaintiffs suffered injury for which the sums set opposite their respective names should be paid."[63]

Each of the ten plaintiffs was entitled to judgment against the United States; the remaining fourteen, as a matter of law, did not show that the government, more likely than not, caused their cancers; therefore, the United States was "entitled to a Judgment of Dismissal as to each of them."[64]

"A legal system," wrote Keaton, "must also provide for growth by occasional instances of more abrupt change."[65] If upheld on appeal, the opinion of federal district court Judge Jenkins will have provided for growth in the law of torts because he took into account a number of factors salient to future petitions by nontraditional, biologically-harmed plaintiffs.

Agreeing with Thode, who wrote that "all conduct is risk-creating,"[66] the Jenkins opinion noted that the federal operatives at the Nevada Test Site owed the plaintiffs the duty of due care and that because there was a lack of due care, the plaintiffs were involuntarily placed at risk. If the defendant's conduct created such health risks for the downwind plaintiffs, and if, subsequent to being placed at risk by the Atomic Energy Commission, identifiable members of that population had health problems, then "they should pay when those risks do result in harm."[67]

While conceding that "the indeterminancy of causation and the long time lag between action and harm"[68] are the characteristics of the dilemma of these unprecedented biological causation tort suits, Jenkins nevertheless moved the law of torts forward using his own "best judgment, experience and common sense." By fashioning a "remedial framework" to examine, individually, each plaintiff's cause of action against the federal government, Jenkins successfully expanded the law of torts. Not afraid to relax the cause-in-fact rules of tort law for these plaintiffs, Jenkins' approach enabled him to rule favorably for ten of the twenty-four plaintiffs in *Allen* on the basis of less than scientifically certain associations between radiation exposure and the subsequent onset of leukemia and certain types of cancer.

In *Allen*, it was impossible for the plaintiffs to show direct proof of causation nor was it possible to show circumstantial proof. In *Allen*, "statistics only enable a plaintiff to prove that the defendant is responsible for an increased number of victims. They do not help the plaintiff prove that he or she is a victim."[69] Jenkins did suggest, given the unusual circumstances of the case, two things: (1) that plaintiffs show relationships between their illness and radiation exposure, and (2) a more dramatic change, that the burden of proof be shifted to the defendant, that is, that the government show why it should not be considered liable for the harms realized by the plaintiffs. What Jenkins did, in effect, was to expand the law of torts by *lowering* "the burden of proof where the defendant's action appears especially reprehensible, so as to allow plaintiffs to recover by showing that causation is possible, or conceivable, rather than probable."[70]

The Response to the Jenkins Decision

The Legal Response

For the plaintiffs and their attorneys the Jenkins decision was a victory, dampened a bit by the fact that fourteen of the plaintiffs were not able to make the case to support the awarding of damages. For Stewart Udall, the Jenkins judgment is a vindication of the system of justice. He urged, again, settlement as the best recourse for reasonable men and women, especially now that the facts and arguments of the parties involved had been revealed through the discovery process. "I have always felt, from the very beginning," he said, after the opinion was announced, "that the best solution, once we had accumulated a lot of powerful evidence, was not a long, drawn-out contest in court, but to have reasonable men sit down and reach a settlement."[71]

His hope is that the government will accept the Jenkins judgment and begin the process of healing the hurts and bitterness that have accumulated since the downwinders first began to suspect the government was not telling the truth. That is probably not going to happen. J. MacArthur Wright, another attorney for the downwind plaintiffs, who practices law in St. George, Utah, said that "the thing that I'm very concerned about is the government indicated it will appeal. I think this is a terrible shame to seek out errors on procedure when it was determined by the court that the government did cause a large percentage of illness."[72]

Dan Bushnell is the attorney for the sheepherder plaintiffs in the *Bulloch* case, which is presently on appeal before the Tenth Circuit U.S. Court of Appeals in Denver, sitting *en banc*. He has argued that since Jenkins has ruled so definitively on the questions of discretionary function, duty, and causation, "the government ought to be spending money on compensation rather than for litigation."[73] The government should bite the bullet, accept responsibility for negligent actions that led to injuries and wrongful deaths, and end the tragedy and the cover-up.

The lawyers for the defendant in the litigation, the federal government, do not share the views of the plaintiffs' attorneys. The Justice Department (speaking for the Reagan Administration), in 1984, was clearly unwilling to provide monetary compensation for the plaintiffs in *Allen*. In an address before the National Association of Attorneys General, held in Washington, D.C. on November 8, 1984, Richard K. Willard, Acting Assistant Attorney General, Civil Division, U.S. Department of Justice, assailed Jenkins' *Allen* decision as "possibly foremost of our common concerns" about the most recent "disturbing development in the law of torts."[74] Willard, speaking on the record for the Reagan Administration, charged that the downwinders "were exposed to very low [levels of radiation], and there is no reliable scientific evidence that those kinds of doses are dangerous at all."[75] (His recorded comments contradicted earlier statements by Administration spokespersons who had testified before congressional committees that the radiation fallout endangered some of the downwinders.)

The Assistant Attorney General, in a simplistic interpretation of the federal judge's legal reasoning, alleged that Jenkins "simply reallocated the burden of proving non-causation-in-fact to the United States as defendant, because otherwise it would be too difficult for the plaintiffs to prove causation."[76] In his address to the assembled State Attorney Generals, Willard insisted that "the district court imposed on the United States the burden of proving non-causation. This means we have to prove a negative, which is often well-nigh impossible."[77] His great concern, and that of the federal government, was that the Jenkins opinion would be "seized upon by other courts which are eager to use tort liability to compensate unfortunate plaintiffs at the expense of 'deep pocket' defendants."[78] He ended his assault on *Allen* by stating, for the first time (although it was not very surprising), "we are appealing the *Allen* decision."[79]

He concluded his conversation with the Attorney Generals by underscoring two essential strategies of the Reagan Administration: (1) "strenuously resisting outrageous new theories of liability and appealing decisions imposing them," and (2) ensuring that, "to the extent we become involved in judicial selection, we need to be aware of the value of judicial restraint, since we are now discovering that even in such a traditional area of common law decision-making as the law of torts, it is surprisingly easy to end up with activist judges engaging in social engineering."[80]

Clearly, the United States' government attorneys are not about to accept Udall's suggestion that they, the Civil Division lawyers in the Department of Justice, sit down with the plaintiffs' attorneys and work out the settlement for the downwind plaintiffs. Willard and his fellow attorneys evidently will spend taxpayer funds to litigate rather than accept, in Willard's words, these "outrageous new theories of liability" created by "activist judges engaging in social engineering."

Legal Parallels

In 1980, a federal interagency task force, looking into the question of compensation for the downwind residents who were injured or killed as a result of radiation fallout exposure, concluded that "there are sufficient parallels between radiation injuries and toxic substance injuries."[81] Many civil cases are pending in federal courts, such as the litigation involving the Nevada Test Site workers, the sheep death case, the uranium miners' cases, the Agent Orange case (just recently settled out of federal district court by the chemical companies and the veterans—but the government is still subject to litigation by both the chemical companies and by the Vietnam veterans and their families), cases involving federal civilian employees at industrial plants such as Rocky Flats, Colorado, and cases involving citizens whose homes were built atop uranium tailing dumps. These cases may be affected by Judge Jenkins' precedent-setting observations about discretionary function and causality. There is also the continuing quest for justice by those among the 250,000 veterans (or their surviving family members) who have developed cancers and/or have seen their children born with birth defects as a result of the radiation they received from participating in the tests conducted by the Department of Defense during the 1950s at the Nevada Test Site.[82]

The plaintiffs' attorneys in the uranium miners' cases will certainly take notice of Jenkins' arguments and try to persuade other federal judges to use the *Allen* case as precedent. Indeed, highly unusual legal actions are involved in the two uranium miners' suits against the federal government. In late

1983 and early 1984, both U.S. District court judges involved in the litigation, Judge Aldon Anderson (Utah) and Judge William Copple (Arizona), stated that they would hold up their final judgments pending the announcements of the Jenkins ruling in the *Allen* litigation.[83]

These two tort actions against the government involve miners, most of whom are native Americans living on uranium-rich reservations, who had worked in unventilated, radiation-contaminated uranium mines and milling plants. The miners charged that they developed lung cancer because of the federal government's negligence in not forcing the private mine and milling plant companies to comply with minimum safety rules.[84]

The Atomic Energy Commission knew about the risk of the miners—who constantly breathed the radioactive gases produced by the uranium dust—developing lung cancer, yet did not take action to prevent illness and death until the 1970s. To keep the price of mining uranium ore low, the AEC "neglected safety."[85]

Mined uranium dust decays into radon, a radioactive gas, and radon, in turn, is broken down into radon "daughters"—the radioactive gas that the miners inhaled—resulting in a dramatic increase in lung cancer. In the 1940s, European producers of uranium ore had installed simple ventilation systems to remove the dangerous gases from their African mines. However, the Atomic Energy Commission "denied that the health problem existed in America. When evidence of this health risk accumulated, it claimed that there was no conclusive proof that radiation caused the lung 'damage' (the AEC avoided calling it 'cancer')."[86]

In the late 1950s Representative (D-California) Chet Holified's Joint Committee on Atomic Energy maintained that there was no danger to the miners. Senator Pete Domenici (R-New Mexico), however, presented shocking medical statistics at a congressional hearing in 1980. His statistics dramatically pointed to the consequences of ignoring the ventilation problem. His presentation included the fact that an Occupational Safety and Health Administration specialist, Dr. Joseph Wagoner, made a 1974 discovery that 144 uranium miners had died from lung cancer in the "Four Corners" area, where the *expected* cancer death rate was only 29.8. Among the 3,500 Indian uranium miners living in this area, there were 114 excess deaths from lung cancer.[87]

In the Utah case, 31 of 111 men who had worked the uranium mines at Marysville, Utah had already succumbed to the ravages of lung cancer when the suit was instituted.[88] The Arizona case involved 85 Navajo uranium miners, stricken with lung cancer, who worked in the unventilated mine shafts on the reservations, and people in the milling plants where the uranium was refined. These shafts and mills were called "cancer factories" by an outraged union official.[89]

Dr. Victor Archer, a physician working for the U.S. National Institute of

Occupational Safety and Health (NIOSH) estimated that of the "original 6,000 uranium miners who worked in the region, 1,000 will die of lung cancer."[90] Since the two federal judges announced that they were waiting for Jenkins' ruling, it is possible that they will adopt Jenkins' discretionary function and causation legal pronouncements and use them as precedents in ruling for the plaintiffs.

Medical Responses

Dr. Charles Land, of the National Cancer Institute, Bethesda, Maryland, argued that the Jenkins opinion was a discretionary one rather than one grounded in the certainty of scientific fact. "A person can use information from science, evidence from science, and testimony from science to arrive at decisions on matters of law," he said, "but they are pretty different."[91] Land, a long-time critic of Lyon's research and his approach to radiation and cancer, had testified as a rebuttal witness for the government in the 1982 trial. Land was uncomfortable with the tentativeness of the scientific assertions employed by Jenkins in the *Allen* decision.

One of Land's colleagues at the Center for Disease Control, Dr. Clark Heath, agreed with Land and concluded that Judge Jenkins "determined that there is enough evidence from a legal standpoint to attach blame. I don't think that is at all a scientific conclusion."[92] Dr. Marvin Rallison, another medical researcher, and a faculty colleague of Dr. Joseph Lyon, has been studying thyroid cancer for two decades. He had studied children years previously and is presently re-examining the same group. After reviewing the Jenkins opinion, he said: "The conclusive scientific evidence is not there."[93]

However, Dr. Joseph Lyon, a leading University of Utah cancer researcher and medical epidemiologist, disagreed—once again—with Land, Heath, and Rallison. Lyon, a major witness for the plaintiffs in the trial, was not surprised that Jenkins had ordered damages in some cases. He observed the care which Jenkins devoted to the scientific issues, and as an epidemiologist who studies the effects of radiation on humans, was pleased with the fact that Jenkins noted that "general range [figures] do not deal with the individuals." "A single community range leaves a lot to be desired," Lyon concluded.[94]

Still others spoke of the impact the opinion would have on the compensatory legislation that has been languishing in Congress since the 96th Congress. While the senior staff person for the minority on the Labor and Human Resources Committee felt it was a fine opinion, Senator Orrin Hatch's staff assistant, Ron Preston, had a different view of Jenkins' opinion. While the Jenkins opinion "provided justice" for the plaintiffs, his less than scientific judgments had thrown a scare into the governmental community, and the senior staff person for the Labor and Human Resources Committee. Jenkins did not work with radiobiologic tables (the essence of the Hatch proposals in 1981 and 1983); consequently, he did not know very much. Jenkins, surmised Preston, "was a fifty-five-year-old judge who doesn't know any science at all and went to law school because he couldn't get into medical school."[95] Surrounded by "friendly witnesses," the federal judge acted on the claims without scientific standards.[96] The response of Congress to the subject of compensatory legislation will be examined in the next chapter.

Mrs. Rula Orton, mother of downwind resident Peggy Orton, probably had the last and most profound word about the impact of the Jenkins judgment. Her daughter died of leukemia at the age of 14, in 1960. The mother, a successful claimant in the *Allen* litigation, received a $250,000 award from the federal judge. Rula said simply: "Of course, money can't pay for a child's death."[97]

Radiation Injuries and the Limits of the FTCA

There will be lengthy appeals of Jenkins' *Allen* decision, all the way to the United States Supreme Court, and it may be upheld by the federal appellate jurists—but this is uncertain. The legal remedy, however, is not the most appropriate one for injured plaintiffs to pursue in this type of negligence action for a number of reasons. First, there is the time involved. The *Allen* litigation began in 1979; the federal trial court judgment was announced in 1984. Five years have elapsed since the legal process began, and it will probably take another three or four years of pending litigation appeals before a definitive remedy is implemented—if one is implemented at all, since the federal appellate court can overturn the federal trial court, and the U.S. Supreme Court can overturn the Tenth Circuit Court of Appeals.

There is a second, more substantial reason why the legal remedy may not be the most equitable one. The common law of torts has evolved to provide justice for individuals when one person has been injured by another "in some concrete, readily identifiable fashion."[98] It is one thing for a federal judge to determine who is at fault in the traditional tort suit in which proximate cause is clearly seen in legal terms by the federal jurist, and the judge, as steward for the people, uses judicial power to compensate the victim, to deter future wrongs, and to provide corrective justice.[99]

But it is quite another matter for a federal judge to face complex scientific questions that require some knowledge of biochemistry, nuclear physics, and medical epidemiology. The nuclear fallout litigation raises critical and complex questions about proximate causality—not only for the present generation of victims and those whose deaths were allegedly due to radiation fallout exposure, but also for unborn plaintiffs (who were in the fetal state during the atomic testing) and the unconceived plaintiffs (future descendants of the downwinders who are born with genetic defects as a result of genetic damage their parents received from atomic test radiation).

U.S. District Court Judge Jenkins, a bright, inquisitive jurist, had asked his lawyers in the *Allen* case to take him down new paths in the area of tort law, and he set precedents as he followed them.[100] However, the more common judicial response to difficult, technical questions associated with cancer, toxic wastes, and other contaminants (carcinogens) and the legal issue of causality is not to stretch the law but to fall back on existing law instead. "Quite naturally, judges faced with complex scientific cases can be expected to gloss over scientific ambiguities in a quest for more familiar analytical models of tort law . . . Increasingly, the substantive determination of tort law requirements, especially legal causation, turns on the court's ability and proclivity for discerning scientific issues."[101]

Considering the serendipitous nature of the litigation route for the downwind litigants and others who claim their serious injuries and their family members' deaths were due to governmental carelessness, i.e., the present-day "disinclination of judges to treat *scientific* proof of causation as germane to *legal* proof,"[102] as well as the fact that it may take time for the tort law to accommodate itself to these new technological wrinkles in the meaning of proximate cause, the person injured by something more exotic than a mail-truck needs to take a more stable path to receive compensation for injuries and suffering.

While the government's own reviewers of the atomic testing controversy,[103] as well as critics of the Federal Tort Claims Act's constraints on the plaintiffs,[104] have discussed modifying the act, both the strong supporters of the plaintiffs' arguments and the government bureaucrats themselves believe that there has to be a better way to seek compensation than through the federal courts. "The proper resolution of this matter may involve a compensation program that is outside of traditional tort law," wrote Favish.[105] "The judicial system is an uncertain source of care and benefits. Neither plaintiffs nor defendants can be assured of the outcome," concluded the Schaffer Report.[106]

From the contemporary perspective, it does seem that a more viable remedy for injury, although very problematical in its own right, is some kind of national legislative compensation plan that would address itself to the problems of proximate causation, over the generations, that have been the stumbling blocks in the litigation dealing with atomic testing and radiation.

The Jenkins opinion in *Allen* is a groundbreaking legal statement. The federal judge, as educator, masterfully took the reader along a path that began with Democritus in ancient Greece and ended with discussions of the double helix, DNA, nuclear physics, and nuclear bomb explosions. There is no doubt that the 200-page opinion will stand as an example of judicial craftsmanship for many decades to come, and that it may very well be the basis for the higher appellate court's modification of the law of torts, sometime down the legal road.

However, possibly its greatest impact now—beyond the award of damages to ten claimants—is its potential influence on cancer compensation litigation in Congress. There has been a fundamental unwillingness on the part of federal executives since President Carter and on the part of Congress, since 1979, to seriously examine the possibility of some kind of political response to the (atomic-testing) tort dilemma.

President Reagan and the congressional leadership cannot help but recall that this grim episode in American history could have been prevented by greater care on the part of the Nevada Test Site personnel. The Jenkins opinion gives the political leadership a factual portrait of agency carelessness that, legally, rose to the level of a willful violation of the rights of the downwind residents and others who involuntarily suffered the consequences of the testing program.

Janet Gordon recently recalled when she began to see a connection between radioactive fallout and her brother's death from cancer. Gordon studied the problems of the uranium workers who were dying from lung cancer. She soon learned about the effects of radiation on sensitive human tissues and cells, and about the onset of cancer. By "educating herself" in this way, she overcame deep feelings of guilt and helplessness over her brother's death, and she began to draw relationships between the atomic fallout and his death from cancer. At no time did the government ever raise the issue of radiation safety with her and her friends in school.[107]

In *Allen*, the Atomic Energy Commission's "casualness is coming back to haunt the government."[108] The Jenkins opinion is a major catalyst in the societal efforts to restore a degree of "fundamental fairness"[109] to this American tragedy reflected so poignantly in the downwind plaintiffs' claims.

"Jenkins' ruling put the stamp of judicial legitimacy on these claims . . . Congress, which thus far has only toyed with the myriad issues raised by government involvement in production, use and regulation of dangerous substances, can play the teasing cat no longer. New standards and limitations to deal with belatedly-recognized, monster threats to citizen well-being must be formulated."[110] It may well be that the Jenkins opinion, with all its future precedential value acknowledged, will have an even greater, and immediate, impact in the halls of Congress and the Oval Office of the White House.

8 The Politics of Nuclear Compensatory Legislation: "Opening Wide the Doors to the U.S. Treasury?"

The nuclear fallout issue is a very emotional one, grossly inaccurate and exaggerated.   President Jimmy Carter, 1979

I am saddened and sickened by the response of the Carter Administration.   Stewart Udall, 1980

The current [nuclear] compensation debate replicates the 1970 era debates over welfare entitlements. The ultimate costs for such a compensation plan could be explosive.
David Stockman, OMB Director, 1984

Allen as the Triggering Mechanism in the Development of a Public Policy

"One of the most politically sensitive issues of the 1980s" will be the compensation of innocent victims exposed to hazardous substances, including radioactive nuclear fallout. So stated a confidential document of the Office of Management and Budget, describing the various congressional "toxic torts" legislation introduced in recent years by concerned legislators. It recommended to President Reagan that he "act now . . . to avoid being stuck with the bills."[1] The victims include the downwinders of southern Utah and Nevada; the Indian uranium miners (more appropriately, their surviving families) who lived in the Four Corners area of the southwest; the Vietnam veterans; the 250,000 military personnel who were exposed to radiation at the Nevada Test Site during the 1950s; the many families living in places such as Colorado Springs, Colorado and Love Canal in New York, who unknowingly built their homes on top of buried toxic wastes and radiation tailings; the tens of thousands of civilian workers who, during World War II, came into daily contact with heavy doses of asbestos while working in military shipyards as plumbers and pipefitters; and the offspring of the injured persons who have suffered genetic damages—all of whom want the federal government to "compensate them for disabilities they allegedly suffer through no fault of their own."[2]

Congress has passed only two such compensatory statutes as of 1984, the Black Lung legislation and the Marshall Islanders "Bravo Test, 1954" legislation, because the legislature acknowledged that environmental agents in the mines and from the atomic bursts in the Pacific had caused adverse health conditions.[3] But the 1984 OMB memo concluded with an ominous view of the consequences that would result from national compensation legislation, including "an open-ended fiscal threat" that would mushroom in cost much like the black lung compensation and social security disability insurance programs have done since their passage in the 1970s. In the spring of 1983, the Reagan White House sent out conciliatory signals about a joint executive-legislative effort to aid the innocent downwinders in Utah, Nevada and Arizona who were victims of government negligence at the Nevada Test Site. However, the OMB memo took another, less positive direction.[4]

Presidents Carter and Reagan clearly refused to support any propitiatory federal response to the plaintiffs' requests for a settlement before and after the *Allen* case came to court. There is also no doubt that the government will appeal the *Allen* case all the way to the United States Supreme Court. Since 1978, OMB directors in both the Democratic and Republican Administrations have strongly encouraged the executive branch, through the Justice Department's attorneys dealing with these litigations, to stonewall the plaintiffs.

In spite of executive hardlining, the Jenkins opinion (and the Agent Orange settlement that was actively encouraged by another federal district court judge, Jack Weinstein) may serve as the *triggering mechanism* that converts a regional problem "into a widely shared, negative public response"[5] toward governmental carelessness and duplicity. This response may lead to a comprehensive, compensatory legislative package or, alternately, to "special case" compensation packages for atomic veterans, Vietnam veterans, and the downwind civilians. The development of public policy in our American polity often springs from judicial opinions which serve as catalysts for change. A judicial opinion such as Jenkins' can become "an obvious conditioning factor and, at times, . . . a critical causal influence on the direction of public policy."[6]

That Americans go to court to resolve their particular grievances is indisputable. In this litigation process public policy issues develop when "individuals with similar problems are forced to cope without solutions for an unacceptable period of time."[7] Occasionally, a Jenkins-type judgment crystallizes the inequities of the issue and focuses on a general grievance shared by many people, not just those who have come into that particular federal court at that particular time in our history. When that occurs, the national legislature is often moved to act.

Jenkins' opinion examines the larger issue of the relationship of an unknowing, trusting citizen and the governmental agent responsible for his or her health and welfare. The opinion focused on the legal, political and normative relationships between the sovereign people and their governing agents, that reach beyond the folks living in St. George and Kanab, Utah, to the larger public. His opinion broadens the conflict that has largely gone unresolved due to governmental intransigence.[8]

People have come to rely on governmental action to change an unfavorable situation—*especially when the government is responsible for their problem.* Apart from its precedential influence on the law of federal torts (should it not be overturned by a higher appellate court), the Jenkins opinion, if used appropriately, could raise the level of consciousness of those in the White House, the national legislators and the larger public beyond the Utah and Nevada borders.

Allen could become the catalyst for developing public policy in the controversial area of federal compensation for victims of governmental operations carelessness and callousness. For the judicial opinion to trigger a change in policy, however, basic economic and political problems must be overcome by the national legislature, and capable congressional leaders are needed for this effort.

Compensatory Legislation in Congress— Some General Problems

Although a number of compensation programs have been introduced in Congress that attempt to remedy injuries and deaths that were environmentally induced by the government or its agents, aside from the Black Lung and Marshall Islanders compensatory legislation (and "Superfund" legislation, to clean-up land and water contaminated by various toxic substances, and preventive legislation or regulations such as OSHA's controversial chemical labelling standards),[9] no programs have been passed by that legislative body.

Almost all radiation victim compensation bills, Agent Orange legislation (and veterans' compensation bills in general), asbestos compensation, toxic substances legislation, and reforms of product liability and workman's compensation legislation have languished in the Congress.[10] In typical "congressional style," the Congress has generally left the business of compensation to the courts—even though the legislators are aware of the terrible burdens they place on plaintiffs, through the limiting jurisdictional and substantive parameters outlined in the Federal Tort Claims Act passed by Congress in

1946.[11] There have been two major exceptions to the legislature's reluctance to legislate compensatory coverage for persons injured due to negligent acts of the government and its agents, including private sector companies doing contract work for the government (such as Kerr-McGee's operation of plutonium plants in Colorado or operation of uranium mines and mills in the Four Corners area).

The Black Lung Benefits Program

The Black Lung compensation program was a "unique congressional undertaking because it singles out for federal attention a specific occupational disease."[12] In 1969, Congress passed legislation that established a compensation system for victims of pneumoconiosis, a chronic, degenerative, generally fatal respiratory disease that afflicts coal miners. The disease is caused by continuously inhaling coal dust, and it leads to painful breathing problems and early death. To qualify for compensation funds, a miner had to demonstrate personal disability due to pneumoconiosis, that the disease arose from employment in an American coal mine, and an employment history of at least 10 years in the coal mines before the respiratory sickness developed.

The afflicted coal miner filed an application with an Administrative Law Judge. For the miners, the process of establishing proof was eased considerably by the specific language of the statute. The bill allowed a claimant to develop a causal link between exposure, disease, and employment. The judge reviewed the factual material, including medical reports and, when relevant, death certificates.

Within a few years of the establishment of the program, over 150,000 miners and over 300,000 dependent survivors were receiving benefits from the black lung program. Until the trust fund concept was incorporated into this 1978 compensation program by a congressional modification of the 1969 statute, the federal government paid all the claims filed by the miners, their dependents, or their survivors, at a cost of over $1 billion a year.[13]

During Congress's 1978 term, the black lung program was liberalized even further and a new financing mechanism was devised to pay for it. Placed on a permanent basis, the 1978 legislation established a tax of 50 cents a ton on underground-mined coal, and 25 cents a ton on surface-mined coal. The tax, which is limited to 2 percent of the sales price in either case, was expected to raise about $100 million during the first year. The tax collections would go into a trust fund, called the Black Lung Disability Trust Fund, to pay for claims filed after 1973 by either the miners themselves or their surviving dependents.[14]

The phenomenal growth of this special compensation bill caused great apprehension in Congress when the issue of compensating nuclear victims was raised. Dr. Ron Preston, Ph.D., Yale University (Medical Sociology), was the senior Majority Staff person on the Senate's Labor and Human Resources Committee, chaired by Senator Orrin Hatch (R-Utah). Dr. Preston believes that the black lung legislation has had a profoundly negative effect on the senators because they saw a modest program develop into an annual billion dollar *entitlement* program, until the 1978 amendments turned it into a trust fund.[15] According to Preston, the Senate feels that "We can't create any more 'Black Lung' type laws anymore."[16]

During his compensation hearings in October, 1981, Hatch had an exchange with Dr. Edward N. Brandt, Jr., Assistant Secretary for Health, U.S. Department of Health and Human Services, about the Black Lung "entitlement" program. When asked by a senator whether the Hatch Radiation Exposure Compensation legislation (S.1483) was an entitlement program similar to the black lung compensation program, Brandt answered: "It is an entitlement program in the sense of those persons who could demonstrate both exposure and the disease process. Yes, it is an entitlement in that sense."

The Chairman, Senator Orrin Hatch, tried to distinguish between black lung and his compensation plan.

> Entitlement programs continue on and on. Black lung has become almost a pension program. It is really a disability program. This [the Cancer Compensation program] is a one-time payment program to pay for the negligence of the government that caused the death or severe injury or illness to people in these three states. Therefore, it would not be an entitlement program. It would not be a continuing program. It is just going to take care of this one problem . . .

Senator Dan Quayle (R-Indiana), responded: "With all due respect, I can recall the same arguments being advanced when we said this was going to be a one-time deal for black lung, and that program has gone on and on and on."[17]

Preston saw the "tremendous proliferation of concern" among senators, as well as the concern expressed by the Reagan Administration, as the fear that the government would be bombarded with toxic tort complaints. This would lead, it was thought, to the breakdown of industries—witness the events surrounding the asbestos litigation and the filing of bankruptcy by the major asbestos manufacturer, the Johns-Manville Corporation, to avoid lawsuits[18] and "other absurd situations."[19]

Marshall Islanders' Compensation Legislation

An equally controversial but incomplete compensatory program addressed the effects of the atomic testing in the Pacific in 1954. During above-ground testing in the 1950s, the AEC set off major thermonuclear bombs in the

Pacific. Due to changed wind currents, one such device, the Bravo Test, blew high levels of radioactive particles over the Marshall Islands. Pressure from the State Department and the islanders resulted in the 1977 enactment of the Marshall Islanders' Compensation Program, which offered the natives $150 million in compensation for the illnesses—especially thyroid cancer and leukemia—and deaths that afflicted the islanders after their exposure to heavy doses of fallout.[20]

On May 24, 1984, the Senate Energy and Natural Resources Committee held two weeks of hearings to examine the Reagan Administration's compact with the Micronesian Islanders. The final settlement of the islanders' radiation fallout claims will provide a $150 million trust fund for the few thousand natives who were living on the island in 1954. After the *Allen* ruling was announced, Senator Hatch's strategy was to make a compact to develop a trust fund of the same amount, $150 million, for the downwinders who developed cancers.[21] However, this compensation program had not, as of the summer of 1985, passed congressional scrutiny.

Typical Federal Disability Statutes

Aside from these two glaring exceptions, which are pieces of legislation that wave red flags for legislators when they hear of compensation programs for victims of toxic carcinogens in the environment, federal law is not very well adapted to the problem of aiding victims of toxic substances, including radiation, contamination and pollution. The only federally financed benefit available to victims of environmentally-caused illness, is federally supported and state- administered disability insurance.

In addition to federal statutes and compacts (Black Lung and Marshall Island Compacts), which contain a causation requirement and an industry-financed trust fund as an integral part of the disability program,[22] there are two congressional approaches to disability. There are legislative statutes for citizens in medical or financial need due to disability that require no proof of causation and follow the traditional tort law requirement.

These statutes, especially Social Security programs, are entitlement programs for persons falling into certain types of categories. Social Security Disability Insurance, which provides income support for persons with long-term disabilities, is the major program in this category. (By 1980 over $13 billion was expended for 2.9 million disabled workers and for 1.9 million dependents.) They were *not* created to compensate victims of disability and illnesses caused by radiation and other toxic substances. However, 53 percent of all the victims of radiation and toxic substances injuries rely on Social

Security Disability Insurance as their sole source of income.[23] Under Social Security Disability Insurance, persons must prove disability only, not the *cause* of the disability. There are also the Medicaid and Medicare programs that provide medical help to persons who have been injured by radiation and by other toxic substances.[24]

Congress also passed disability statutes that follow the traditional tort law by requiring persons to identify the substance and party responsible for the injuries that have become disabilities.[25] Statutes that follow this general pattern include the Federal Employee Liability Act, 1906 (railroad employees and seamen); the Federal Longshoreman's and Harbor Worker's Compensation Act, 1926; the Price-Anderson Act, 1957 (nuclear industry workers' protection legislation); the Federal Water Pollution Control Act; and Veteran's compensation legislation.

All of these statutes require the use of the "preponderance of evidence" standard before disability claims will be awarded for injuries and wrongful deaths.[26] Most have restrictive clauses that make it difficult for victims of toxic substance injuries, including radiation exposure, to receive compensation. For example, veterans must show that their illness is service-connected and that it manifested itself within one year of the end of their active military service. (In the case of cancer and leukemia, there is no way to meet that stringent requirement unless Congress changes the VA regulations; the 98th Congress made dramatic changes in that direction.)

With the exception of the black lung and Marshall Islander legislation (the latter legislation not yet law), the proposed nuclear victim radiation compensation legislation, and the changes in federal laws affecting veterans, present federal law does not provide individuals injured by toxic substances with anything more than "limited and haphazard compensation."[27] Congress is unwilling to open up the Pandora's Box of compensation for victims of toxic substance and nuclear radiation injury for a number of reasons.

"The cost of compensation bills would run into the billions, either for industry or for the federal government," wrote one observer of the congressional scene in 1983.[28] Historically, the remedy for the downwinders' illness and disability caused by the nuclear testing has been state workman's compensation, Social Security Disability Insurance, or a legal remedy through tort liability actions in federal courts. All these approaches to the problem were tightly controlled by stiff eligibility requirements. The recent compensation bills proposed in Congress are of a different order of financial magnitude.

In the asbestos litigation alone, there have been over 24,000 claims filed against asbestos manufacturers for their failure to warn the users about the health hazards of asbestos. As of 1983, 3,800 were settled but at an astronomical cost: $400 million in compensation claims, $164 million in legal fees, and $565 million for the manufacturers' legal fees![29]

Congress has been considering asbestos compensation legislation since 1973, *six years before* the introduction of Senator Edward Kennedy's Nuclear Compensation Legislation in 1979. However, no action has been taken. Moreover, the largest asbestos manufacturer, Johns-Manville, filed for bankruptcy to avoid the probability of receiving 74,000 cancer claims and 177,000 asbestosis claims between 1980 and the year 2009.[30]

At the Love Canal, where homes were built atop toxic chemical wastes, "rough estimates for clean-up . . . are $30 million to $40 million, while residents are reportedly seeking as much as $2 billion in damages" (health and disability as well as property).[31] In response to these statistics, even sympathetic congresspersons such as Representative Bob Eckhardt (D-Texas) have said that "in a perfect society, we would take care of victims of disease. But if we can't even provide hospitalization and medical care, we can hardly agree to compensate" on a general basis.[32]

The costs of special compensation programs for victims who have been exposed to radiation, asbestos, and other toxic substances, are being weighed by the President and Congress at a time of unprecedented growth of the federal deficit. For example, welfare and income maintenance programs—entitlement programs such as Social Security Disability Insurance, social security, medicare, food stamps, Aid for Families with Dependent Children—have jumped "a staggering 600% in the decade ending 1981."[33] While the Reagan Administration has made an effort to reduce the number of individual (categorical) programs from 57 to 9 block-grant (general) programs, with an accompanying 25% reduction in expenditures, the politics of social welfare programming in Congress has enabled a number of the programs to keep going strong. The food stamp program has increased from 19 million participants receiving $7 billion in 1979 to 23 million persons receiving $9.7 billion in 1981.[34]

Thus, the compensation bills for victims of radiation exposure come to Congress at a time when both the President and the Congress are trying to "trim federal expenditures and get a handle on the mounting federal deficit."[35] Not only are the costs of fixing people's lives so high, but the legislators are wary of having a compensation program for the thousands of persons who were living downwind of the Nevada Test Site turn into a continuing, disruptive, draining entitlement program for other groups in the larger society who have also been injured and wrongfully killed due to government negligence. The "black lung" phenomenon haunts many legislators and makes them unwilling to go along with what Hatch terms a one-shot remedy for the downwinders.

In addition to the heavy monetary burden that the radiation exposure compensation legislation carries with it when it enters Congress, there is also the equally heavy evidentiary burden. Legislators are extremely wary of meddling in areas where the scientific community has not definitively answered the question of causality. The fact that federal district court Judge Jenkins has ruled on the issue has made the evidentiary issue only slightly less onerous for legislators. "We as legislators ought to leave the decision up to the scientists," said Representative David E. Satterfield III (D-Virginia), when asked whether specific environmental factors, such as the nuclear fallout from the above-ground atomic tests at the Nevada Test Site, were the cause of the cancers and leukemias that developed in the downwind communities.[36]

The two major Senate protagonists in the nuclear victim's compensation legislation, Senators Ted Kennedy (D-Massachusetts) and Orrin Hatch (R-Utah), differed fundamentally in the evidentiary methodology used in their versions of the compensation legislation. Senator Hatch's legislative assistant on this policy issue, Dr. Ronald Preston, was firmly committed to a legislative methodology that used radiobiological tables (developed by the U.S. Department of Health and Human Services) to statistically determine whether or not a person's cancer or leukemia might have been caused by radiation exposure.

Preston was convinced that only a screening method based on scientific notions of "attributable risk" and "probability of causation" would convince legislators that the Hatch bill was more than a "black lung" entitlement program. He maintained that his senator's approach to compensating the victims of radiation, which emphasizes an *"objective, scientific standard"* to assess the probability of a person's cancer having been caused by the fallout, was ultimately the only method that would be acceptable to the gun-shy legislators.[37] (He was, as events turned out in the 98th Congress, wrong.)

The *event-specific* approach to compensation was used in the 1979 Kennedy legislation and is still considered to be the most appropriate strategy for passing a compensation bill for the victims of the above-ground atomic tests. Paralleling Jenkins' opinion, the Kennedy bill presumed that the government was negligent, that it acted wrongfully, and that the people living in the communities downwind of the Nevada Test Site developed cancers and leukemias because they were placed involuntarily at risk by the federal government in the 1950s.[38]

Republican staffers criticized this *event specific* legislative strategy as cost-inefficient, and reminiscent of the black lung legislation that rapidly grew out of control. However, the minority staffers insisted that the Kennedy approach was less politically volatile than the Hatch bill because the Kennedy plan was limited only to the atomic veterans and the citizens living downwind of the Nevada Test Site.[39] Although a consensus has developed among medical epidemiologists researching the association between cancer and low doses of radiation, and federal and state judges have ruled in favor of the plaintiffs' survivors in Colorado and Utah (and settlements have been reached in New York and Wyoming on associational issues), legislators are still pushed and pulled on the question of evidence.

The evidentiary problem was well summarized by a staff person on the Senate Committee on Labor and Human Resources. Michael L. Goldberg stated recently: "One minute a guy is operating a punch press and he has five fingers, and the next minute he has four. You know when, where, how, and why it happened. But cancer? Did he get it from benzene, or the time he worked in a pesticide plant, or from smoking, or because his father had it?"[40]

In the public policy debate over compensation for victims of government's carelessness there is little pressure group support for this type of legislation except from the affected groups themselves. Environmental groups such as the Sierra Club have always pushed extremely hard for legislative funding for *clean-up* of the environment and the *prevention* of nuclear disasters or hazardous waste tragedies, but they have not supported compensation for victims of nuclear accidents, negligence, etc.[41]

In addition, the various constituencies adversely affected by these negligent governmental actions have only recently begun to form networks with each other. It was only in 1983 that Citizen's Call leader Janet Gordon was able to develop a nuclear fallout roundtable of leaders of various groups who were separately pursuing litigation against the federal government for injury and wrongful death due to governmental negligence.[42]

Legislative supporters of the compensation plan have also had difficulty developing these networks because, in the words of Republican staffer Preston, many people see the legislation as sort of a pork barrel "private bill," like the Hatch bill, that will benefit only the people of southern Utah.[43]

Conversely, a number of effective lobbying groups have opposed this type of compensatory legislation. The most active opponents of the legislation are the industrial lobbyists (chemical industry, nuclear power industry), insurance companies (who must provide the plaintiffs/claimants with the funding that flows after settlements and after monetary awards and who "present stiff opposition"),[44] and most importantly, the federal executive branch bureaucracy. For both minority and majority staff persons on the Labor and Human Resources Committee in the Senate, the organized "non-partisan" opposition by Office of Management and Budget officials, Department of Energy, Department of Defense, Department of Justice, and White House staff—from Carter through Reagan—have worked very effectively to kill the compensation legislation.[45]

One Republican staffer recalled an ominous meeting, in April 1982, with employees of the Department of Defense and the Department of Energy, who brought him to a dilapidated room in a deserted building in Washington and said, "We have to stop this [compensation] project. The National defense requires that you stop it . . . This legislation will demolish nuclear safety standards and cost the government untold billions." The staff assistant quickly walked away, and they did not harass him again.[46]

"Only absolute opposition" from the "bureaucratic establishment in the federal executive branch," has been present since 1979, stated Dr. Preston. Using the "Chicken Little" argument again and again, i.e., that the sky will fall if Congress passes a radiation compensation bill, top officials in the Veterans Administration, Office of Management and Budget, and the Departments of Energy, Defense, and Justice have worked long and hard to prevent the legislation from getting out of the committee system and have stonewalled plaintiffs' attorneys' efforts to reach settlements in the nuclear radiation litigation.[47]

Despite these obstacles—cost factors, evidentiary problems, lack of legislative support for the nuclear victim legislation, and a very effective opposition, especially from the Executive Branch of the federal government—the compensation legislation for victims of atomic fallout was initially developed in the late 1970s in Congress.

The Nuclear Victims' Compensation Legislation, 1979–1984

The 1979 Legislation, S.1865

Senator Ted Kennedy's 1979 legislation, the Radiation Exposure Compensation Act of 1979 (S.1865), was an effort to amend the Federal Tort Claims Act to make the United States liable for damages to certain individuals who lived in the downwind area between January 1, 1951 and October 31, 1958, or between June 30, 1962 and July 31, 1962; to certain uranium miners who worked in a uranium mine in Colorado, New Mexico, Arizona, or Utah for a year between January 1, 1947 and December 31, 1961, and to certain sheep herds, damaged by certain nuclear tests at the Nevada Test Site in these time periods.[48] As presented in 1979, the bill presumed absolute governmental liability for damages and wrongful death to persons and survivors whose deceased parents or children resided in the geographic areas outlined in the legislation, and who developed specified illnesses:

> There shall be an *irrebuttable presumption* that the damages alleged by the plaintiff were caused by exposure to radiation as a result of a nuclear detonation or exposure to uranium as a result of employment in a uranium mine, as the case may be.[49]

With the "irrebuttable presumption" clause embodied in the statute, an eligible person had only to file a claim in accordance with the Federal Tort Claims Act guidelines; however, since there would be no need to show negligence and causation, the court would simply determine the damages that were to be assessed. The tort laws of Utah or Nevada or Arizona would apply in these cases, since the Kennedy legislation did not set upper limits on the amount of the damage award.

The result of the passage of the Kennedy legislation, according to a hostile Carter Administration Task Force report issued in 1980, was that "the government would be required to pay substantial money damages to all individuals who developed specified forms of cancer" [leukemia, thyroid cancer, bone cancer, or any other cancer identified by the Advisory Panel on the Health Effects of Exposure to Radiation and Uranium] . . . [50] notwithstanding the fact that very few of these could be attributed to radiation."[51]

Hearings were held on this controversial "event-specific" Kennedy legislation, strongly supported by Hatch (who was on the minority side of the aisle when the legislation was introduced.) In joint hearings sponsored by the Senate Judiciary Committee and the Senate Labor and Human Resources, Subcommittee on Health, the senators heard from an array of Carter Administration officials who argued strongly against the bill.

The two major protagonists that day, June 10, 1979, were Stewart Udall, chief attorney for the downwind claimants then awaiting a response from the Department of Energy, and William G. Schaffer, special litigation counsel, Civil Division, Department of Justice.

Schaffer testified first, after hearing the testimony of a number of Navajo uranium miners and their families. He had two major criticisms of the Kennedy legislation. First, that it was "extremely overbroad in application," and "would provide damages to anyone who resided within the specified geographic area during the period of testing and who subsequently contracted one of a number of forms of cancer."[52] Second, the spokesperson for the Carter Administration maintained that S.1865 was "inappropriate and would place an enormous and unwarranted burden on the American taxpayer."[53]

Regarding the potentially huge cost to taxpayers, Schaffer's 1980 report to the President was highly critical of the Kennedy legislation. If passed, he wrote, it would have an "undesirable impact on other federally funded programs and [there was] the possibility that it would become a model for the resolution of other toxic substances problems," for example, those of veterans who took part in military exercises at the Nevada Test Site and of the civilian workers at the site.[54]

The 1979 Kennedy legislation, warned Schaffer, "could appreciably affect the operation of the VA's multi-billion dollar disability compensation program" and could have an adverse impact on the toxic torts issues that were surfacing in Congress and in the federal courts—dioxin (Agent Orange), asbestos, benzene products, cotton dust, and beryllium.[55] There were, Schaffer stated, "sufficient parallels between radiation injuries and toxic substances injuries," to create major concern within the Carter administration about the ultimate costs.[56]

Schaffer argued, instead, for an approach to the problem that Hatch's legislative assistant Preston (himself a medical epidemiologist) urged the Senator to take later, in 1981. The Justice Department official called for "the identification and compensation of those people who may have been injured by exposure to radiation from the atmospheric weapons testing program," but warned that the task would be "extremely complex and perplexing." "In all candor," he conceded, "we have not yet arrived at an acceptable solution."[57]

In 1979, however, Hatch was still a defender of the Kennedy strategy and he criticized Schaffer on a number of occasions, especially questioning Schaffer's impartiality as Chairman of an Interagency Task Force looking into the association between cancer and nuclear fallout, while he was, at the same time, in charge of 120 Department of Justice attorneys handling over 900 claims from cancer claimants.[58] At one point in the heated exchange between the Senator and government bureaucrat, Hatch exclaimed: "You and I both know that statistics and computer models can make [dose information] come out anyway you want with regard to these matters." To which Schaffer replied: "I don't think that is really right."[59]

After the government witness departed, a solemn Stewart Udall took the witness stand and indicated how "saddened and sickened" he was "by the response of the Carter Administration here today."[60] What sickened Udall more than the intransigence of the Carter administration on the question of settlement for the downwind claimants was the imperious attitude of the federal bureaucrats. For example, after Schaffer's admission that, "yes, we probably did hurt some people,"[61] Udall heard the Department of Justice official excitedly state that "the testing was particularly dramatic . . . It is a particularly dramatic form of technological hazard."[62]

Udall was a strong supporter of the Kennedy legislation, because in effect it would have required the federal courts to waive the cause-in-fact requirement and to assume governmental negligence and wrongdoing when the federal judges heard radiation exposure cases under the Federal Tort Claims Act. Udall, however, was realistic enough to know that the bill would not see the light of day without strong executive backing.

The Carter Administration was not in support of the legislation, and no further action was taken by the 96th Congress. By the time the next bill came before the Senate, the Senate was in the hands of the Republican Party, and Orrin Hatch (R-Utah) had taken over the chairmanship of the Senate Labor and Human Resources Committee from Edward Kennedy (D-Massachusetts).

By 1981, moreover, Senator Hatch had been persuaded by Dr. Ron Preston, his Labor and Human Resources legislative assistant, to move to a new, radical legislative strategy for the revised 1981 nuclear compensation legislation. After 1981, Senator Kennedy and his staff were in the minority on the committee and consequently with the minority with respect to the strategic development of the radiation exposure compensation legislation. This shift in legislative planning proved to be the downfall of Hatch's efforts, during the 97th and 98th Congresses, to pass legislation that would benefit the downwinders living with the bitter legacy of governmental carelessness.

Also, from the beginning of Hatch's chairmanship of the Labor and Human Resources Committee in 1981, he had little control of it. Two Republican senators, Robert T. Stafford (Vermont) and Lowell P. Weicker (Connecticut), carried strong labor support in their re-election battles and moved away from Hatch's position on key issues before the Committee. Without their votes, and confronted with the Democratic leadership of Ted Kennedy, Hatch was actually in the minority (7–9) on key bills.[63]

As Chair of the committee that dealt with the nuclear victims' compensation legislation, Hatch was not terribly successful; "on many of his former causes he had either lost or had declined to fight at all."[64] Since "Hatch did not command a majority on his own committee," and because he "did not command as much respect as he does attention,"[65] the compensation program did not move very far along in the political process. Hatch was seen as a "bit of a fanatic" by many of his senatorial colleagues, and as something less than a "heavy hitter" by those on his committee.[66] His role as marshaller of forces for the compensation legislation was a very problematic one.

The 1981 Legislation, S.1483

According to Senator Hatch's senior staff assistant on the Labor and Human Resources Committee, the Utah Senator had come to dislike the methodology in the Kennedy legislation. For Hatch, the 1979 Kennedy bill "was like the black lung legislation," that is, the *event–specific* legislation drew circles, relative to location, event, and cancer, and if a person fell into a circle, funding would be available for injury and wrongful death due to the irrebuttable presumption that the government was the cause of these illnesses and deaths.[67] Preston was very critical of the "presumptive standard" inherent in the bill and was able to persuade the Senator to move away from this *event-specific* type of cancer compensation legislation.[68] The bill that Senator Hatch introduced early in the 97th Congress was based on a strategy of determining associations between nuclear testing and cancer on a set of radiologic data tables developed by medical researchers.

The Hatch proposal's primary methodological distinction was to take radiation data—amount of dosage received, cancers typically radiogenic in nature, statistical tables of the probability of developing leukemia or cancer based on certain continuous exposures—and to relate it to individuals who claimed that their injuries and (in the case of surviving spouses' allegations) deaths were due to the fallout from nuclear testing or from the "radon daughter" gases in the uranium mines.

Hatch's 1981 bill, jointly offered to both the Senate Judiciary Committee and the Senate Labor and Human Resources Committee, was an effort to modify the Federal Tort Claims Act to enable the downwind plaintiffs to receive a remedy for governmental negligence. While it established the liability of the government and generally followed the time, geographic and disease criteria of the Kennedy bill, it indicated that "if the plaintiff meets such requirements, there shall be a *rebuttable presumption* that damages alleged were caused by government activity at the NTS."[69]

This was a very important distinction. Unlike the 1979 Kennedy legislation which emphasized the *irrebuttable presumption* standard, the Hatch proposal was written so that the presumption that the government's negligence caused the illness could be rebutted in the federal district court by the government attorneys—but the burden of proof was on the government to show that it was not at fault.

The bill also provided for the use of radioepidemiological tables to protect against nonmeritorious claims. If there were at least a 10 percent probability that the person contracted cancer or leukemia due to radiation exposure from the atomic bursts or from the gases in the nonventilated uranium mine shafts, then a certain amount of compensation—on a one-time basis—would be made available. The monetary range was limited to actual damages awards of between $50,000 to $500,000 per person. The Department of Health and Human Services was responsible for developing these probability tables and an *Advisory Panel on the Health Effects of Exposure to Radiation and Uranium* was also to be created if the legislation became public law. Its task would be to continue to refine these statistical tables and identify other types of radiogenic cancer that could be placed on the list used in the compensation awards process.

If a person lived in the affected areas during a certain period of time (from 1951 to 1958 and 1962), and contracted a certain type of cancer readily associated with radiation exposure, then based upon statistical tables and medically developed probabilities, there would be a monetary award for actual damages with an upper limit of $500,000. This procedure also governed uranium miners who worked in the Four Corners area between 1947 and 1961 and who contracted lung cancer. Additionally, the Hatch bill would have provided a remedy for the sheepherders who lost livestock after the 1953 Nancy and Dirty Harry atomic blasts.[70]

Hearings were held on the Hatch legislation in October, 1981, and again in the Spring of 1982. The Reagan administration sent a phalanx of critics to the October, 1981 session. No less than thirteen bureaucrats from the

Department of Health and Human Services, National Institutes of Health, Defense Nuclear Agency, Department of Defense, Department of Energy, National Cancer Institute, Veterans Administration, and the Environmental Protection Agency attended the public hearing. Their obvious goal was to torpedo the Hatch legislation. While the Reagan people did "recognize" that there was exposure to radiation produced by the fallout (not a single Reagan Administration person addressed the tragedy of the uranium miners), and that "some of these individuals may have suffered adverse health effects from this exposure,"[71] they completely rejected the Hatch proposal.

Additionally, Hatch had to try to educate some of the senators on his committee. Senator Dan Quayle (R-Indiana) asked if there were any way to determine which individuals contracted cancer from the radiation exposure. Hatch responded quickly: "You can, on the basis of probabilities and on the basis of legal proof."[72] However, the government officials rejected this type of proof as incomplete and medically and scientifically unsound.

Dr. Edward N. Brandt, Assistant Secretary, Department of Health and Human Services, Interagency Radiation Research Committee, saw the bill as an entitlement act. (He also did not know where St. George, Utah was, an embarrassing moment for the head of a research group that allegedly had focused on this area for its radiation exposure studies. In response to a question from Senator Paula Hawkins (R-Florida) about research activities in the St. George area, Brandt replied, "I am sorry. I do not know where the St. George area is.")[73]

The most candid witness for the Reagan Administration was Lieutenant General Harry A. Griffith, U.S. Army, Director, Defense Nuclear Agency, Department of Defense. While admitting that some people might have been injured by the fallout from the tests, he argued that (1) Hatch's bill was not supported by facts; (2) the methodological procedures it would have the federal judge employ were inappropriate and unworkable; and (3) it *would severely restrict current and future nuclear programs and activities in the United States.*

> Perhaps the greatest problem associated with the bill as we see it is the potential effect on all activities involving millions of workers (who have received the same exposures as the downwind persons) and the use of ionizing radiation and radioactive substances. As you know, sir, our nation is heavily dependent upon these activities which necessarily involve the exposure of workers to low levels of ionizing radiation. Examples of this would include research, medical and dental procedures, nuclear power, industrial radiography, nuclear weapons, and nuclear weapons testing.[74]

In addition to his major concern about the legislation's impact on the continued use and development of nuclear power, Griffith was concerned about its cost and about the many "ineligibles" who would be receiving compensation. By amending the Federal Tort Claims Act to remove the burden of proving negligence and causation from the plaintiffs' shoulders, the Hatch proposal increased the likelihood of unjustified compensation, maintained the General.[75]

The manager of the Nevada Test Site, Mahlon E. Gates, put this government position quite bluntly when he said to Hatch's committee, in response to their allegation of government negligence: "All right, but I believe that the first man who comes forward and says 'my cancer is caused by you guys who put out radiation' is obliged to show causation and to show negligence on the part of government *rather than our just handing out money as this bill purports to do.*"[76]

As in 1979, Stewart Udall appeared before the committee in 1981 to discuss the proposed legislation. Wearily, the lawyer defended the legislation (with some reservations): "I feel a little bit like a hardened battle veteran now."[77] Once again, Udall condemned the government's intransigence on this compensation question. After the outcry in 1979, he recounted, the "administrative agencies hunkered down and waited for time to elapse."[78] Udall found General Griffth's testimony refreshing because he was quite candid in his effort to protect the nuclear establishment in the United States. However, he described the other "administration boys" as using "squid tactics" and "trying to obfuscate the issues again," by using techniques such as averaging. Said Udall indignantly: "There is no human being named St. George . . . Human beings got the doses; They [the government] will not study the human beings."[79] But, saying he was "past anger in all this,"[80] Udall focused on the Hatch legislation and offered his concerns about the bill.

Udall liked the area-specific approach of the Kennedy proposal and felt that the Hatch bill stood little chance of passage.[81] He also took issue with the Hatch bill's limits on monetary awards, the exclusion of unreimbursed medical expenses, the limitations on the forms of cancers that were allowable under the Hatch proposal, and the makeup of the Department of Health and Human Services advisory panel.[82] He perceived, however, that Hatch was committed to using the radiobiological tables. Knowing the political drawbacks that such an approach would have, Udall went back to presenting the plaintiffs' case in the federal district court in Utah later that summer.

In March, 1982, the Reagan Administration put out tentative feelers to Senator Hatch's staff, indicating that they were "ready to work" with the Congress on a compensation plan. In March, Brandt, the Department of Health and Human Services bureaucrat, called Hatch's bill the "best approach" to the problem. Hatch, extremely optimistic at that point, claimed that this Reagan turnaround was a "light in the darkness" compared to earlier Carter and Reagan Administration statements on compensation.[83] (Though the feelers went out in March, 1982, by the end of April, Hatch complained that the administration had not yet officially endorsed the bill.)[84]

In the spring of 1982, additional hearings were held by Hatch's committee. Governor Scott Matheson (D-Utah) was critical of the Hatch proposal. He recommended a bill in which the government would "admit liability and . . . let the courts decide causation and recovery."[85] Other victims of radiation exposure testified before Senator Hatch's Committee in Salt Lake City, on April 8, 1982. These citizens strongly supported the area-specific legislation but would have settled for the Hatch bill as a last resort.

On April 20, 1982, the Labor and Human Resources Committee, chaired by Hatch, voted 14 to 1, to report S.1483. The sole dissenting vote was cast by Oklahoma Senator (R) Don Nickles who believed that it would open up the Treasury of the United States to persons who had developed cancers unrelated to the radiation exposure. The next stop for the bill was the Senate Judiciary Committee.

On May 6, 1982, it was sent to the Judiciary Committee. Senator Hatch firmly believed that he could get the bill marked up and onto the floor of the Senate for action before the July 4, 1982, holiday recess. Hatch claimed that chances for full Senate approval were "extremely high" but, he cautioned, "it's a real race against the calendar."[86]

His bill didn't make it past Senator Strom Thurmond's (R-South Carolina) Judiciary Committee. It died there, bogged down by strong senatorial opposition, presidential opposition, and the strong lobbying efforts of the insurance industry.

In the final analysis, there was very little Republican support in the Senate for the legislation.[87] Furthermore, the Reagan Administration never actually supported the Hatch bill during the spring of 1982. The Reagan Administration opposed the bill because of the huge potential costs of a settlement, its devastating effect on the United States' nuclear industry, and its potentially damaging impact on the Reaganomics strategy. The Reagan Administration was also concerned about the impact of such legislation on the government's nuclear global defense strategies.[88] (Buttressing this executive position, the Congressional Budget Office came up with an estimate that it would cost $14 billion alone to compensate victims of just one type of cancer.)[89]

The Democrats in the Senate were opposed to the legislation as it emerged in 1981 because it was "Hatch's" bill. Further, the insurance lobby actively attacked the legislation. James L. Kimble, senior counsel for the American Insurance Association, argued that the Hatch bill would lead to new pressures to adopt S.1483 for other toxic tort controversies.[90]

The 97th Congress ended its deliberations without a compensation bill. The Senate members of the Labor and Human Resources Committee, especially Kennedy and Hatch, were at odds over the basis of the compensatory legislation itself. In the 98th Congress, Hatch, once again, introduced the compensation legislation, but this time with even more restrictions.

The 1983 Legislation, S.921

The senator introduced his legislation, *Radiogenic Cancer Compensation Act of 1983*, in late March 1983. It was referred to the Senate Committee on the Judiciary—not his Committee on Labor and Human Resources. Unlike the earlier bills introduced during the 96th and 97th Congresses, Hatch was the sole sponsor. *Not a single legislator joined him in sponsoring the cancer compensation legislation.*

Kennedy and Hatch continued to disagree over the bill's methodology. A senior staff person for the Democratic senator stated that there was "overwhelming opposition" to Hatch's legislation and that "the bill ground to a screeching halt."[91] From its inception in the 98th Congress, "there was no political movement" on the cancer compensation bill at all.[92] There were no hearings scheduled on the proposal. It was introduced and then left to die after being referred to the Subcommittee on Administrative Practice and Procedure.

Even Preston, the senior staff person for the Republican Party members on the Labor and Human Resources Committee on this issue, admitted, in the spring of 1984, that "S.921 is a dead letter."[93] While there were some "passive collaborators" such as Senators Dennis DeConcini (D-Arizona) and Paul Laxalt (R-Nevada), there was a pervasive "general neglect" in the legislature.[94]

If the 1979 proposal had the *irrebuttable presumption* standard, and the 1981 legislation had the narrower *rebuttable presumption* standard, the 1983 proposal did not have this "presumptive" clause at all. Instead, it used the tort liability "standard of proof," stating that "for purposes of this Act, an individual must demonstrate his individual dose by a preponderence of the evidence."[95] Except for this substantive change, which lessened the bill's impact a great deal, Hatch's legislation was very much like the 1981 proposal. The two major aspects of the 1983 legislation, according to Hatch, were the use of radioepidemiological tables and formulas and the provision that allowed damages to be paid to plaintiffs whose probability of harm from the fallout or from the "radon daughters" gas was as low as 10 percent.

Hatch had clearly committed himself to the statistical approach to the problem of association between cancers and nuclear fallout. He was unwilling to make the presumption, *in the law*, that the government was at fault and that the only thing left was for a judge to determine damage awards to plaintiffs. The *event-specific* approach of Kennedy—and of Utah Governor Scott Matheson and Judge Bruce S. Jenkins in the *Allen* case—was anathema to the conservative Republican. As Preston put it, "Hatch was not taking a 'Jane Fonda' approach to the problem."[96]

His 1983 proposed legislation extended to the atomic veterans and to the uranium miners, most of whom were American Indians. Under the proposed legislation, a civil action (under the Federal Tort Claims Act) could be instituted in a federal district court against the United States, by or on behalf of individuals (or the estates of individuals) who had contracted a radiation-related cancer diagnosed after January 1, 1952 (or January 1, 1948 for the uranium miners).

To file suit against the government for its *proven* damages against individuals who claimed cancer injury or death due to government testing, the bill provided for the use of probability tables developed by the Department of Health and Human Services of the federal government. If a probability of causation existed on the basis of statistics and probabilities, then actual damages awards could be made from $50,000 to $500,000 by the federal judge.

For Scott Matheson, Utah's Governor, the Hatch bill was not a good one because the remedy for the illness or death was a predetermined, limited one.

> I never felt good about Hatch's bill. He was trying to squeeze down on the recovery. I never felt that was legitimate. I felt that in a democratic society, when you make a mistake you got to stand up and you got to pay for it—the traditional way through a tort suit.

The Governor believed that Congress "should establish a statutory confession—that the government was wrong and that it is willing to pay for its indiscretions—and then, with causation and negligence determined, have the plaintiffs go back to Jenkins for the proper relief." Matheson somberly admitted that he had "frustrated Hatch" but said that Hatch had frustrated him too in his effort to achieve a measure of justice for the victims of governmental negligence at the Nevada Test Site.[97]

Opposition to the Hatch proposal continued to come from the federal government. The Reagan Administration, still extremely concerned about the financial impact of compensatory legislation and "really worried about domestic perceptions of radiation,"[98] continued to play hardball with the senator. For the Administration and for Hatch's opponents in the legislature, Hatch's 1983 version of the bill was the dilemma. A senior staff person said that "the bill cannot prevail" with a mechanism that could be employed by other constituencies to claim governmental negligence and subsequent injury due to lack of due care by the government or its operatives.[99]

The minority staff person argued that people working in the nuclear power industry, radiologists, people living downwind of nuclear power plants or nuclear research facilities, and others who have had contact with ionizing radiation or some other toxic substance, have become very concerned about associations between ionizing radiation and cancers. With the knowledge made available through the Hatch bill, the possibility of getting cancer would become "far more frightening to a lot more people."[100]

For Carter and Reagan administrators, the bill raised the specter of the entire nuclear industry coming to a halt because of the inordinate fears of tens of thousands of workers and millions of people living near nuclear plants and research facilities. The Reagan Administration, like the Carter staff, was extremely concerned about the grave influence publishing the Hatch proposal's radiobiological tables would have.

The Jenkins *Allen* opinion, clearly an event-specific approach to the dilemma of the downwinders, was another indirect rejection of S.921. The federal judge presumed, after reviewing the evidence on the record, lack of due care—negligence—by the federal government's operatives at the Nevada Test Site. Jenkins then proceeded to evaluate the claims on a case-by-case basis. The minority staff counsel for Labor and Human Resources believed that Jenkins' opinion supported their compensation strategy. The Hatch staff, however, saw the opinion as bad law—judicial policy-making at its worst. The Jenkins opinion was, in Hatch's words before a congressional committee, in May, 1984, "legislation by judicial decree."[101]

Hatch was in a quandary. Given the massive indifference and, worse, hostility in the Congress, the extreme opposition of the Reagan Administration, and the "stiff competition of the insurance companies,"[102] there was little he could do to sell S.921 in the 98th Congress. The Utah senator, however, devised another strategy for assisting the downwinders in the three states adjacent to the Nevada Test Site. He tried unsuccessfully to add an amendment to the compact negotiated in 1983 between the Marshall Islanders in Micronesia and the United States.

In 1983, Ambassador Fred M. Zeder II, with the U.S. Department of Interior's Office for Micronesian Status Negotiations, concluded a *Compact of Free Association* with the Marshall Island leaders. In the compact, in which the Islands, an American trust since 1947, would be recognized as an independent nation, the government settled all radiation injury claims of the Marshall Islanders (about 3,000 of whom were affected by the atomic and hydrogen bomb testing) by creating a $150 million health care trust fund—but without admitting the United States' culpability.[103]

On May 24, 1984, when the Senate Energy and Natural Resources Committee began hearings on the proposed compact, Hatch appeared and called for the committee to establish, through an amendment to the compact, a similar $150 million health care trust fund for the downwinders in Utah, Nevada, and Arizona. Claiming that it was unfair to establish a trust fund for the islanders while not providing a similar package for Americans suffering the same bitter legacy from atomic bursts, Hatch urged the Committee to provide the equity for the citizens downwind of the Nevada Test Site.

In light of the fact that a federal district court judge has ruled that the federal government was negligent and that the Nevada tests did cause cancers among American citizens, and in light of the fact that numerous congressional hearings have led most of the members of Congress who attended them to a similar conclusion, and in light of the fact that this administration seeks to establish a $150 million trust fund to settle the Marshallese claims that they and their property were damaged by our Pacific nuclear tests, do you not agree that it makes sense to establish a comparable trust fund to settle the comparable claims of American citizens?[104]

Zeder's response to Hatch's rhetorical question was that the compact was a matter to be judged separately from the domestic situation. The compact "served American security interests and was needed to reduce friction between the United States and the Marshall Islands."[105] Asked by the Utah Senator to explain why the Americans shouldn't be treated on the same terms as the Marshall Islanders, a deputy assistant Department of Defense Undersecretary, James Kelly, said, "I can't give you a reason."[106]

Hatch's basic strategy in the 98th Congress was extremely ad hoc. Given the lifelessness of S.921, due to Hatch's own intransigence, all he could do for his constituents was to propose a $150 million health care trust fund in the form of an amendment to the Marshall Island Compact, to be considered by the full Senate, probably in the summer of 1984. In this effort, said a minority staffer, Hatch "may be successful."[107] He was not successful.

"Life is uncertain," said Preston, Hatch's assistant on Labor and Human Resources.[108] While S.921 was born without a chance of success, the ad hoc amendment to the Marshall Island was seen briefly as the vehicle for some kind of remedy for the downwinders in the three southwestern states.

The *Allen* Opinion, Congress and the Future of Cancer Compensation

Immediately following Jenkins' *Allen* ruling, editorials in the *Deseret News* and the *Salt Lake Tribune* speculated on the impact it might have in Congress. "Fallout Verdict Can Prod Congress To Take Action," suggested the *Deseret News*, while the *Tribune*'s editorial was entitled, "Ruling Underlines Dire Need For New Damage Guidelines."[109] Congress, these comments clearly suggested, must address the question of a *remedy*. While Jenkins' opinion "struck a blow for justice and accountability,"[110] the editorials anticipated interminable delays due to the appeals process. Because of the grave fear that such a ruling would open a "Pandora's Box of demands for huge payments by people claiming they were affected by chemicals, medical and industrial radiation, and the like,"[111] the government will appeal at every level in the federal appellate process. However, both papers called on the Utah congressional delegation to take some kind of action in order for the Congress "to produce some fair plan to compensate those families that have endured sickness and death from the nuclear test years."[112]

Governor Scott Matheson, the outgoing Democrat in the Utah state capital, urged the Utah congressional leaders to seize the opportunity afforded by the *Allen* judgment and resolve the downwinders' nightmarish legal dilemma. Congress should be encouraged, he told the press, to enact legislation exposing government culpability in order to settle the thousands of claims that still had to be heard in federal court. "I would prefer to see Congress make the policy decision about liability. There's no reason to try the whole series of cases on negligence."[113] Let the legislation, by admitting government liability, "leave Jenkins the responsibility of awarding damages to all fallout victims without further legal action."[114]

The specific strategy he proposed fell on deaf ears in the Utah congressional delegation. Two Republican congressmen, James V. Hansen and Howard C. Nielsen, and Senator Jake Garn remained silent on the issue. Congressman Dan Marriott (R-Utah), a lame-duck representative (he had announced his retirement from Congress in the winter of 1984), argued that the judgment "gives us third-party credibility." He believed that the opinion would enable the House Judiciary Committee to reexamine the issue of compensation for victims of the testing (through a revision of the Federal Tort Claims Act), and he pledged to "strike while the poker's hot."[115] Marriott's inability to act decisively on these issues was recognized, however;[116] the compensatory legislation had to come from senatorial initiatives taken in the Senate. And in the Senate, the leading figure was Senator Orrin Hatch.

In the course of the six-year debate on this issue, Hatch had lost the support of all 14 co-sponsors of his legislation because of his unwillingness to work cooperatively with other legislators in his campaign against the lobbyists for the industry, insurance companies, and the White House. Instead of using the Jenkins opinion as a triggering mechanism to legislate a remedy, Hatch and his staff *criticized* the opinion. Preston even derided Jenkins as a frustrated, 55-year-old who really wanted to be a doctor but wasn't good enough for the profession![117]

Tragically, at a moment when a judicial opinion was developed that could have been used to marshal support from other senators because of the character of the decision itself, the person who had the opportunity to dramatically act on the matter did not. Hatch could have provided both a realistic short-term remedy for the downwinders and a long-term remedy for the victims of negligence with toxic substances. However, Hatch was ideologically wedded to a methodology that scared legislators, industry, insurance companies, and the White House. While *Allen* "increased the pressure on Congress and the President to settle with the fallout victims,"[118] and while the sentiment for settlement may have been present after the Jenkins ruling, Hatch's seeming ineptitude quickly closed the window for action on the matter of justice for the downwinders.

This was not the case for the veterans who had been exposed to radiation and toxic substances such as the dioxin in Agent Orange. This constituency had capable legislative movers in the Congress. During the 98th Congress, while Hatch's bill was stymied, Senator Allan Cranston's (D-California) legislation, *Veterans Dioxin and Radiation Exposure Compensation Standards Act of 1983* (S.1651), was introduced on July 20, 1983. It called for a "presumption of service connection" to be established by the Administrator, Veterans Administration, for certain diseases of certain veterans exposed to radiation or dioxin during military service. It made provisions for resolving claims from the veterans or their estates and was placed on the Senate calendar for full debate in late November, 1983.

Meanwhile, the House of Representatives passed a similar bill (HR 1961) by voice vote on January 30, 1984. The bill, *Agent Orange and Atomic Veterans Relief Act,* authorized disability, compensation or survivors' benefits for Vietnam veterans who suffered or died from diseases associated with Agent Orange.

As reported out of the Veterans Comittee on January 25, 1984, by a vote of 30 to 0, the bill also contained a provision for compensating "atomic veterans," i.e., personnel (or their surviving families) who were exposed to ionizing radiation during the above-ground atomic bursts in the 1950s or during the occupation of Hiroshima or Nagasaki after 1945, who subsequently developed leukemia or thyroid cancers. On January 31, 1984, the House-passed bill was sent to the Senate for its consideration when it reviewed the Cranston legislation.

The May 7, 1984 Agent Orange settlement, engineered by federal district court judge Jack Weinstein, spurred further congressional action according to Congressman Thomas Daschle (D-South Dakota), himself a Vietnam veteran and the sponsor of the House bill. After the settlement was announced, Daschle said, "if the companies can do it, then the government can act as well."[119]

The House bill had 205 co-sponsors. None of the three Utah congressmen co-sponsored this legislation which was so closely related to crucial issues for the downwinders! It is a tragedy that the three Utah representatives could not see the relationship between the atomic veteran and the civilian downwinders.

Without a congressional policy initiative that responds to the downwinders' injuries, separate trials for the plaintiffs will prove to be a lengthy legal battle, and a morally tragic one for society. Hatch's inability to marshal support in the Senate for such policy modification, given the twin legal, political, and moral supports—the Agent Orange settlement in a New York federal district court and Jenkins' *Allen* judgment in a Utah federal district court, in the same week—speaks to the personal style and political capability of the Utah legislator.

While the two federal judges, Weinstein and Jenkins, provided the legal and normative basis for the legislative effort to resolve the injustice, there had to be a moving force in the legislature to capitalize on these judicial decisions. Unfortunately for the downwind civilians, that force was not present in the 98th Congress.

Unlike the military veterans who were exposed during the 1950s to the same fallout from the same bombs which burst over the Nevada Test Site in the early morning hours, the downwinders did not have representatives in the 98th Congress with the legislative acumen to capitalize on the federal judge's strong opinion. If a remedy does develop in the Congress, and that is the place for the development of a comprehensive program that would provide the downwinders with distributive justice, it will be established despite the Utah delegation's inability to seize an opportunity to move the legislators to consider a public policy for a societal problem that, according to some medical observers, has yet to emerge in its complete horror.

9 Rights and Remedies for the Downwinders

Ultimately, our society is unlikely to tolerate a situation in which harm is continuously suggested.   Dr. Ronald Preston, 1983

All of us, in a social sense, are the cause of the problem, either as beneficiaries or as victims.   Judge Charles D. Breitel, 1983

The "Dark Side of Technology"

The twentieth century has undergone, and is still undergoing, a staggering technological revolution. "Chemistry, physics, medical science and research, electronics and mechanics have conjoined in a bewildering multitude of uses" that have led to "a profusion of affluence of nutrition, services, medication, transportation, clothing, shelter and countless other benefits regarded as positive and even essential goods."[1] Many of this century's technological and scientific breakthroughs, especially harnessing the atom, were "the acknowledged offspring of an official union of science and government."[2]

However, these benefits of science and technology were accompanied by risks exemplified by the atomic bomb and the activities of the government and scientists.[3] The atomic testing program at the Nevada Test Site involved two distinct types of risks. The obvious risk was the danger of the ionizing radiation produced by the detonation of the atomic bombs. The second risk was the consequences of introducing "more effective governmental control over the uses of science in public planning and decision making."[4]

Both sets of negative consequences of the atomic bomb's development, (1) the effects of radioactive fallout, and (2) the government's coercion of local agricultural agents and pathologists and the continuous lying of its agents, such that citizens were "unable to distinguish truthful messages from deceptive ones,"[5] lay bare, in the words of President Lyndon Johnson, "the dark side of technology."[6]

The dark side of the nuclear technology began to emerge in 1945. With the desert detonation of the first atom bomb, at the Trinity Site in White Sands, New Mexico, in the summer of 1945, "science outgrew its age of innocence."[7] "Now we are all sons of bitches," exclaimed one observer of the Trinity blast. Little did that onlooker understand the terrible risks that

195

millions of persons would have to confront, without their informed consent, after the 1945 blast.

Only in recent years have "the terrible costs of this technological revolution in disease and death . . . been uncovered."[8] The lengthy, bitter tragedy of the downwinders is symptomatic of the larger social, political, and moral dilemma our political system confronts: How to compensate the many injured and dead persons that technological advances have left in their wake.

In the 97th and 98th Congresses, concerned policymakers introduced compensatory legislation that would deal with some of the presently recognized, adverse consequences of the technological revolution. The dark side of technology includes such consequences as toxic chemicals in the workplace, asbestos exposure, black lung injuries, brown lung injuries, Agent Orange, oil spill injuries, and hazardous waste dumps (in 1983 alone, the EPA concluded that 70 *billion* gallons of toxic wastes had been illegally dumped by industrial concerns).[9]

This list also includes atomic veterans who received high doses of ionizing radiation from improperly conducted atomic tests at the Nevada Test Site in Camp Desert Rock, and the downwinders who were continuously exposed to atomic blast radiation.[10] During the 98th Congress in March, 1983, the Congressional Office of Technological Assessment published a major study that expressed its concern about these radiation-related injuries and deaths. In addition, the congressional office recommended that in the future, government–science/technology programs write into the legislation or regulation "assurance that in the event of damages to health or environment, or indirect economic effects, injured parties will be able to obtain equitable compensation *expeditiously*"[11] [my italics].

There is no question that terrible costs have been incurred as a result of the technological revolution. Equally unimpeachable is the fact that "our present mixed tort-administrative system for compensating persons has serious major deficiencies."[12] Changes are needed in national legislation which would modify the tort system or devise a more equitable administrative remedy for persons injured as a consequence of new technological developments. It may be that the compensatory packages created by Congress for persons injured or killed due to government negligence may be extremely costly to society and have an adverse impact on our economic system.[13]

However, immediate remedy must be provided for the injuried persons and for families of persons who have died as a result of governmental carelessness, negligence, and duplicity at the Nevada Test Site during the aboveground atomic testing period. It is now clear that "responsible persons at the operational level of continental nuclear testing neglected the important, basic idea: There is nothing wrong with telling the American people the-truth."[14] After nearly thirty years of negligence, duplicity, and coverup by governmental operatives who implemented the above-ground atomic testing program in the Nevada desert in the 1950s, it is time to deal with this dark episode in American history. Too many people needlessly suffered great pain and died because of governmental deceit and a shocking lack of care by Nevada Test Site scientists and operatives, to avoid a substantive principled response by the government that wronged them.

The Imperative of a Remedy for the Downwinders

By the time the above-ground tests began in 1951, government scientists and technicians on-site *knew* the dangers of ionizing radiation.[15] "Scientific knowledge about both the pathological and genetic hazards of ionizing radiation was substantial," said Favish, in 1950.[16] We now know that during the 1950s the AEC had implemented safety procedures at its own laboratories Lawrence Livermore, Los Alamos, and Oak Ridge—to protect its lab personnel from the dangerous effects of radiation.

But no protection was afforded the downwinders, who were given no warning and were told that exposure to ionizing radiation from the atomic detonations posed no threat to their health. The governmental operatives minimized the dangers of fallout. The 1953 detonation of "Dirty Harry," for example, which dropped heavy accumulations of radiation particles on St. George and other downwind communities, was made light of or covered up by nervous AEC operatives.[17] No adequate dosimetry readings were taken or kept by the AEC during this time, readings which would have determined the actual doses received by the St. George residents. The AEC speculated that only a small amount of rads fell on the town after the Dirty Harry detonation, with subsequent fallout of no more than 5 rads in a small number of hot spots. Not only were the downwinders misinformed about their safety, they were trapped in a bureaucratic operation that prevented the truth from emerging about the actual levels of radiation exposure. Three decades after the radiation exposures, the medical researchers' preliminary conclusions now indicate that the radiation dosages in St. George, Utah were probably underestimated by 400%! University of Utah radiobiologists used the thermoluminescence method to examine samples of brick from St. George. They determined that the doses *were from 19 to 20 rads,* using one process, and estimated as much as 20 and 22 rads, using another radiobiological process.[18]

The record in this unfortunate episode in American history is replete with numerous examples of governmental operational duplicity, harrassment, cover-up, and betrayal of the trust of patriotic citizens in order to continue the continental testing program. Consistent with conceptions of fairness, the government must try to rectify these past errors by swiftly compensating for the injuries, illnesses, and deaths that occurred as a result of the improper, unconstitutional and unethical behavior of federal officials.

The atomic testing issue, like Watergate, reaffirms the fact that no one in government is above the constitutional requirements of procedural fairness and fundmental due process. During the atomic testing program in the continental United States, between 1951 and 1963, federal government agents intentionally withheld basic information from tens of thousands of United States citizens. They prevented the publication of medical research studies that would have threatened the continued operation of their elaborate bureaucratic activity, and they harrassed and coerced local doctors and public health officials into silence when local residents grew concerned about the dangers of exposure to radiation from the atomic blasts.

The downwind citizens, concerned about the ash-colored radiation flakes that fell on their loved ones, homes, gardens, and playgrounds, were ridiculed by the governmental agents—the same agents who took great care to protect themselves from accidental exposure to nuclear fallout. The residents of the Utah, Nevada, and Arizona towns downwind of the testing site were seen by the federal operatives as uneducated yokels who could be buffaloed and placated with slick public relations gambits.

These residents, unknowingly placed at risk by federal government operatives at the Nevada Test Site and by AEC managers in Washington, D.C., need to be made whole again—psychologically and medically—by the government that has outraged them. Formal confirmation by the government of the downwinders' beliefs about the AEC's negligence, carelessness, lying and cover-up (and of federal judge Jenkins' discoveries about the government's Nevada Test Site activities—which influenced his decision in the recent *Allen* litigation), would help in the recovery of these people's larger sense of purpose and of a world-view which was betrayed. Compensation for those who have suffered and for the families of those who have died, is right and just. The remedy must be found by the government that wronged the downwinders.

Compensatory Remedies for the Downwinders

There are essentially three potential public remedies available to the downwinders: the political remedy, the administrative remedy, and the presently used (and partially successful) judicial remedy. The Congress, with the Administration's cooperation, could pass legislation to *create a new administrative agency* that would provide remedial relief for the downwinders. Alternately, the Congress could *pass legislation to modify the role of existing federal agencies, such as the Department of Health and Human Services or the*

Veterans Administration, and make them responsible for dealing with the downwinders and other groups that were needlessly and involuntarily placed at risk by the national government in the 1950s. Finally, the Congress, in admitting governmental culpability, *could amend the Federal Tort Claims Act to enable the federal district court judges* to quickly and positively respond to the question of a remedy for the downwinders.

To act in any of these three ways, the various congressional factions in both houses of Congress must coalesce on (1) the need to deal with the problem and (2) the strategy. In the creation of a successful domestic public policy, it is equally critical that the Executive branch play an integral part in the process of constructing legislative contours with legislators (through political activity by the President's White House and agency surrogates). Also, there is always the need, in formulating a viable public policy, for strong cohesive support from various groups outside the formal channels of public policymaking.

Obviously crucial to the resolution of the national tragedy that occurred at the Nevada Test Site are the views and actions of the two political branches of our national government. Unfortunately, the views of both the Executive and Legislative branches have been extremely cautious and fragmented, and their actions have been clearly negative. Equally dismal for the downwinders' cause has been the reaction of environmental and other public interest pressure groups who have not been involved in this particular policymaking process. These groups have left a vacuum to be filled by other pressure groups, such as insurance companies, who have been opposed to any kind of compensatory public policy.

While finding a cancer cure has been, since the Nixon Administration, a "prime topic of national concern" in the Congress and the White House,[19] the question of compensating persons whose cancers arose from governmental negligence is a much more problematic one. Political actors and environmental pressure groups have focused on prospective dilemmas associated with leukemias, other cancers, and toxic substances, and have steadfastly refused to look backward to those who were injured or killed by governmental inefficiency and error.

The Executive branch of government, since the Carter Administration, has dramatically resisted all efforts by the downwinders' attorneys to settle the victims' complaints. Their resistance is broadly based and present in (1) administrative (Department of Energy and Veterans Administration) rejections of petitions for redress by claimants, (2) unwillingness by government attorneys in the Departments of Justice and Energy to even discuss settlement during the discovery process in civil litigation, and (3) absolute rejection of numerous legislative proposals presented since 1979 by concerned senators and representatives who attempted to provide a remedy to those persons irreparably injured by governmental carelessness and deceit.

Federal agencies that are research-based, such as the National Cancer Institutes, Veterans Administration, and National Institutes of Health, and the federal agencies involved in regulating the nuclear testing program (Defense Nuclear Agency, Department of Defense, Nuclear Regulatory Commission, and the Department of Energy), have followed the White House lead regarding the question of an association between nuclear fallout and cancer. Lawyers for the government, who represent the agencies in administrative hearings or in federal district court litigation, repeatedly maintained the "no-association" argument—even in the face of growing statistical and epidemiological evidence that there were highly significant associations between the fallout and the ensuing cancers and leukemias.

For example, Department of Justice attorneys working with the *Allen* litigation in the Utah federal district court continuously denied that the government had been negligent, when all the world could see—by viewing the government's own secret AEC documents—a clear pattern of deceit and lack of due care by federal operatives at the Nevada Test Site. In the last two years, the Veterans Administration has only reluctantly come to accept some sort of association between the government's activities and the citizens' illness, *but only for purposes of medically treating military veterans at VA hopsitals.*

Given the decision-making process in the legislature, i.e., the fragmentation of power and decentralization of authority down to the subcommittee level, attempts to develop a logical policy for remedying the downwinders plight has been extremely difficult. In Congress, "there is no single such policy" either for a remedy or for the larger problem of dealing with technological disaster to prevent future cancers and leukemias.[20]

"Rather than resolving legislative ambiguities and stalemates, . . . Congress has often abdicated its authority."[21] Except for legislation attempting to redress millions of Vietnam and atomic veterans, Congress has moved slowly into this area of public policy. Beyond the legislative dilemma of "poor congressional leadership and factional discipline,"[22] another hinderance in developing compensatory legislation has been that the various presidential Office of Management and Budget directors, among others, have been joined in battle against remedial legislation *before* the bills emerged from congressional subcommittees.

Given (1) a fragmented and cautious Congress that is lacking effective legislative leadership on the question of remedy for the downwinders; (2) a phalanx of White House bureaucrats, from the President and Office of Management and Budget directors downward, who vigorously attack the compensation plans; and finally, (3) the fact that both Democratic and Republican Presidents, Carter and Reagan, and three Congresses (96th through 98th) have been greatly concerned (and therefore unwilling to act) about the potentially staggering costs of a compensation program for the downwinders (and others), the judicial remedy has been the only one availabe for possible equity for the downwinders.

However, until the Jenkins judgment in the recent *Allen* opinion, still subject to the uncertain and lengthy federal appellate review, the federal judicial remedy had not been a viable one for the downwinders. Before U.S. District Court Judge Bruce S. Jenkins broadened the scope of the traditional Federal Tort Claims Act remedy, claimants who had gone into federal court seeking a legal remedy for governmental negligence in radiation litigation had been unsuccessful in convincing federal jurists of associations that existed between governmental negligence and injury. But the Jenkins decision *could* be overturned on the merits of a higher appellate tribunal. Should that occur, the downwind petitioners and their attorneys would be back to square one—eight or more years later, and with even more bitterness toward the political system. Technically, the *Allen* opinion is limited in its value as a precedent until such time that a higher appellate court accepts Jenkins' interpretation of the Federal Tort Claims Act as the consensus of the Tenth Circuit or of the U.S. Supreme Court. Unless that occurs, Jenkins' legal ruling must be treated as a narrow one, applicable in a single jurisdiction and, possibly, for a very limited period of time.

The Foreseeable Future for the Downwinders

The short-term prospects for the downwinders and their attorneys are not optimistic. Small groups such as Citizen's Call and the Downwinders, perpetually short on funds, have had difficulty informing the community about the dilemmas that confront the downwinders. There is little community money available for Janet Gordon's venture to attend a conference on radiation victims in Washington, D.C., and no funds to establish necessary communications between these volunteers and the downwinders in southern Utah, Nevada, and northern Arizona. Citizen's Call activities seem to fall on the shoulders of one energetic and highly motivated woman, Janet Gordon, who has to wait on tables when the funds run down.

The picture is a bleak one when one reviews recent compensatory legislation for the downwinders. Since 1979, it has languished in the legislature, unsupported by legislators and roundly attacked by White House bureaucrats who have periodically visited the Capital to lobby against the Kennedy and Hatch proposals. While there was a slight chance of adding an amendment to the Marshall Islands Compact in 1984 to create a trust fund for the American downwinders, the legislative picture is generally discouraging for the survivors. It is especially poignant that a national dilemma has been turned into a local one, perceived by legislators as Hatch's "pork-barrel" bill for a few Mormons who live in southern Utah.

The Reagan White House, committed to reducing nonessential, government financial support in the domestic arena, has completely opposed any kind of legislation that would provide taxpayer funds for persons injured by negligent governmental activities in the past. Reaganomics is based on reducing entitlement programs, and the Reagan personnel see a compensation bill as a type of entitlement that would commit the federal government to an expenditure of untold billions of dollars to these and other victims of the government's lack of due care. Consequently, it has made every effort to kill the legislative effort and has been unwilling to discuss a settlement with the plaintiffs' attorneys in the *Allen* litigation.

"The Pink-Orange Clouds of Dust"

The downwinders' best and probably only hope for the foreseeable future, therefore, is the judicial remedy. Despite the very nature of the federal judicial system, with its lengthy time delays and prohibitive costs, the plaintiffs have the ability, at least in federal court, to mount a substantive challenge to governmental negligence—within the limits of the existing law. Until the political branches reevaluate the complexities of the downwinders' problem and develop a public policy that addresses society's responsibility to react to governmental wrongs committed against its citizens, the only remedy is the one found in the federal district courts.

The plaintiffs have found only one compassionate ear in the government—that of the federal judge who, having heard the testimony and having read the documents, concluded that these people had been lied to and deceived by government operatives. How *Allen* will be received by the federal appellate judges, on review, is a matter of conjecture.

It is important for the appellate judges to understand the thirty years of history of the Nevada Test Site litigation and to understand the clients who "sought solace for their test-blamed sorrow."[23] It is equally important for the appellate judges reviewing the *Allen* record to understand the two duties inherent in the downwinder litigation in Utah's federal district court. Jenkins wrote that first "the law imposes a duty on everyone to avoid acts in their nature dangerous to the lives of others."[24] The second duty, for Judge Jenkins, "was the duty of judgment in this case," the judge's duty "to provide justice."[25]

In carrying out his judicial duty, Judge Jenkins found a gross lack of due care on the part of the government operatives during the entire period of above-ground atomic testing at the Nevada Test Site. For over 12 years, the AEC operatives did not act to avoid needlessly endangering the lives of the downwinders. If stringent safety standards were in place at the government's AEC nuclear research labs, then why was not "at least as much care exercised in dealing with the . . . radioactivity in the pink-orange clouds of dust, gases, and ash drifting eastward from southern Nevada?" asked the federal judge in the *Allen* opinion.[26]

The "ethical rule, the moral tenet" was ignored by the government, and because of this fundamental denial of due care by governmental agents, injuries and deaths were more likely than not to have occurred in ten of the cases that came before the federal judge in the *Allen* litigation.

Jenkins had stared at the "dark side" of technology and concluded that the government was under a legal and moral obligation to pay for the damages inflicted upon people who were unknowingly and involuntarily placed at risk by the federal government. It is also critical for our society's stability for the political system to resolve, once and for all, the terrible consequences of the AEC and Public Health Service bureaucrats having recklessly implemented the continental atomic testing program at the Nevada Test Site over thirty years ago.

"Innocent victims of the technological revolution should be compensated," wrote Judge Breitel.[27] The downwinders living in southern Utah, northwest Arizona, and southeast Nevada were the innocent victims of an uncaring federal bureaucracy composed, in part, of knowledgeable university scientists employed by the federal government. Government accountability must be established for the wrongs it committed against the thousands of persons who trusted and believed in it.

Judge Jenkins' judgment, if not overturned, is a sound beginning for the hard, yet necessary, re-evaluation of the federal government's formal response to the administrative, legal and political petitions brought by the downwinders. It is hoped that, sooner rather than later, the federal government's political branches will modify their positions on the issue of compensation for the victims of radiation fallout. Given the public record to date, replete with numerous documents that uncover duplicity, deception, and lack of concern about the downwinders, the President and Congress must get to the task of developing a public policy that takes the right and just position on an issue of ethical and practical importance to the downwinders and to the larger community. Congress and the White House, abiding by the general constraints of the Constitution, are also committed to providing just remedies to felt wrongs. The people who lived downwind from the Nevada Test Site have been greviously wronged.

Postscript

On October 9, 1946, Colonel Stafford Warren, the officer in charge of radiological safety at Operation Crossroads, America's Pacific Ocean atomic testing program, wrote a memo to General Groves about the insidious danger of fallout. On November 12, 1985,[1] in Federal District Court in California, it appeared in a civil suit. The Justice Department stated[2] that "government officials and scientists were aware of the hazards of radiation since the inception of the nuclear weapons programs, . . . specifically that fallout could cause cancer."

Warren's memo to Groves was one of many documents presented to show that, from the beginning of the testing program, the U.S. government knew that radiation exposure *would* kill people:

> [A radioactive fragment from a bomb casing] is probably the most toxic metal known, and . . . extremely small amounts deposited in the marrow will eventually cause progressive anemia and death years later. Tumor formation has a high incidence. . . . [The material] mixed with these fission products, beta and gamma emitters, is an insidious hazard—not immediately dangerous but if absorbed into the body it produces a long time hazard. . . . The amount necessary to cause this hazard is minute—measured in millionths of a gram. The harmful effects occur years later. . . . *I believe a frank statement of this sort should be made now to professional and intelligent lay groups as part of the general discussion on the effect of the bomb as a whole.* . . .[3] [my italics]

Another government document released in the *Broudy* litigation, Colonel Warren's 1946 lecture to radiological safety workers (eerily reminiscent of Gloria Gregerson's recollections of her exposure to fallout as a child), underscores the knowledge the government had of the probable consequences of exposure to fallout:

> You need only to absorb a few micrograms . . . to develop a progressive anemia or a tumor in from 5 to 15 years. This is an insidious hazard and an insidious lethal effect hard to guard against. . . . [Radioactive fallout would be] all around you, . . . you couldn't eliminate it and it would get on your clothes, in your house, in the water, in the milk, and all the food. It would be in the dust and in the air you breathe. Filters couldn't keep it out. . . . You get it on your hands, you transfer it to your bread and jam, and you ingest it. You pile up the amount—although it is not readily absorbed you gradually pile up increasing amounts. . . .[4]

And eventually, people would die of leukemia or cancer. "Simply put," said the Department of Justice in *Broudy*, "the government *knew* of the hazards of radiation"[5] in January, 1951. But (rejecting Warren's suggestion) AEC agents *deliberately* chose not to share that knowledge. They chose, instead, to equate radioactive fallout with dental x-rays. The radioactive fallout tragedy began with less than a noble lie[6]—and continues to unfold thirty-five years later.

Appendix A: The Long Journey of *Bulloch v. United States,* 1953–   : Polaris Sighted?

It was said by Frost that justice is the opposite of mercy—so is one end of a transit opposite the other; but through it Polaris may be sighted and the course of justice set.

"The Line of Light," from *Judicial Afterthoughts,*
A. Sherman Christensen, 1974 (Courtesy, Ken Verdoia)

In 1953, the AEC reported that many thousands of sheep were grazing in an area close to the site of the two dirty shots that took place during the spring, 1953 Upshot-Knothole series, Nancy (24.4 kilotons) and Harry. Very soon thereafter, in the spring and summer of 1953, 25 percent of the new lambs and 12 percent of the ewes died.

A preliminary AEC study, confidential until 1978, found "extraordinary concentrations" of radioactive iodine–131 in the bodies of the dead sheep. AEC scientists, Drs. Arthur H. Wolff and Monroe A. Holmes, were then taken off the investigation and replaced by others whose conclusions ignored the autopsy results.[1] (Also ignored were 1951 AEC reports warning that years after the 1945 Trinity atomic test, cattle and sheep were still showing the effects of radiation burns and of having ingested radioactive vegetation, which led to deformities in the animals' offspring.)[2]

In a July 15, 1953 meeting the AEC commissioners heard from one of their own medical staff, Dr. John Bugher, that after the Dirty Harry test it had been "clearly established that a number of horses had been injured by beta radiation and that the AEC liability was probable." Burgher reported that "the situation was not so clear" with the sheep, and concluded that "probably their deaths had resulted from the ingestion of toxic plants; however, since they apparently also suffered some radiation injury, studies would be continued."[3] However, for public consumption, the AEC refused to admit that the sheep were killed by massive doses of radiation.

Reviewing AEC reports on this incident in 1963, Dr. Harold Knapp was shocked to find that the AEC did not mention the large amounts of *internal* irradiation of the dead lambs' thyroid glands (20,000 to 40,000 rads) and the terrible damage to the dead sheep's gastrointestinal tract (between 1,500 to 6,000 rads) from "all the fission products present in the fallout particles which were ingested along with the open range forage."[4]

However, Paul Pearson, M.D., Chief of the Biological Branch Division of the AEC, told an Iron County, Utah agricultural agent that the AEC "could under no circumstances afford to have a claim established against them and have that precedent set."[5] On February 17, 1955, a complaint by local ranchers against the AEC for the deaths of over 17,000 sheep was filed in the federal district court in Salt Lake City.[6]

Bulloch v. United States, 145 Supp 824 (1956)

In the 1956 case before Federal District Court Judge Sherman Christensen, who was selected randomly by lot for these cases, ranchers from Southern Utah brought suit under the provisions of the Federal Tort Claims Act against the federal government for damages because of thousands of sheep allegedly killed as a result of the radioactive fallout from the atomic bomb tests in the vicinity of the range where the sheep were pastured.

Evidence produced by the government indicated that the sheep deaths were not caused by the radioactive fallout. Instead, local AEC doctors contended that the losses were:

> the natural result of unprecedented cold weather during the lambing and shearing of sheep, inadequate feeding, unfavorable winter range conditions, and infectious diseases of various types.

After reviewing the "law phases" of the case, i.e. the "discretionary function" exception, and drawing the basic distinction between policymakers and those who administer the policy, Judge Christensen, on the factual record presented to the Court in 1956, concluded that the sheep deaths were not caused by the atomic radiation.

While Christensen found that the AEC had been negligent, i.e., the AEC operators had had "the duty to use reasonable care to ascertain [civilians'] whereabouts within areas to be affected and to at least give them timely warning" and there "were no advance warnings given or other precautions taken to safeguard the herders and their sheep," the evidence in 1956 seemed to indicate that the radioactivity did not kill the animals.

Christensen accepted the "overwhelming weight of the expert testimony marshalled by the government to demonstrate that point . . ."[7] "It does not lie with the Court to question the great weight of the testimony that these differences were not determinative, in the absence of at least some evidence that they were . . . The great weight of the evidence indicates, and I find, that the doses were well within the permissible maximums for human or animal body tolerance."[8]

The federal judge, while very concerned about the AEC administrators' evident negligence, nevertheless refrained ("it does not lie with the Court to question the great weight of the testimony") from ruling in the plaintiffs' favor because he felt he had received

> positive testimony from those best in a position to know that the maximum amount of radioactive fallout in any area in which sheep could have been, would have caused no damage.[9]

Because the testimony and evidence presented by the federal administrators did not link the sheep's deaths to the radiation fallout, and, as Christensen stated, he refused to question the AEC scientists' "integrity,"[10] he dismissed the action on the merits. He did not question the quality of the evidence presented nor did he suspect that the AEC administrators would be involved in deceptive practices before the federal judge. "The great weight of the evidence" led Judge Christensen to rule against the plaintiffs. Even Dan Bushnell, counsel for the plaintiffs, accepted these findings and stated, after Christensen's decision, that "counsel for plaintiffs is willing to waive the findings of fact on this matter, in feeling that the court's analysis is sufficient for them. And under due deliberation, we have no intention of appealing this or offering any counter proposal."[11]

Bulloch v. United States, 95 Rules Dec 123 (1982)

Following (1) the 1979 congressional hearings that uncovered evidence of the AEC's duplicity in the above-ground atomic testing program, and (2) the testimony before a congressional committee of Peter Libassi (General Counsel for Carter's Department of Health, Education and Welfare) and Dr. Donald Frederickson (Director of the National Institutes of Health) that the sheep were irradiated (during which time Libassi stated that "dose levels were almost 1,000 times the permissible levels" for human beings),[12] the same plaintiffs asked the Federal District Court to set aside the earlier judgment and to re-examine the charge on the merits. (Governor Scott Matheson (D-Utah), after reviewing, in 1979, the documents on Dirty Harry and the sheep's deaths, had exclaimed: "This is the most blatant example of instance upon instance of official evasion and manipulation of facts about the health impacts of fallout . . . The air of secrecy and agency-self protection that pervades the documents made careful, scientific inquiry impossible.")[13]

In August, 1982, Judge Christensen, who had heard the 1956 case, determined that there had been "conduct by the government that amounted to a species of fraud upon this court" in 1956, and he set aside the prior judgment.[14]

At the trial, held May 10 to 13, 1982, three basic sets of evidence were produced by plaintiffs in an effort to show fraud and negate the the 1956 judgment: (1) The AEC had knowledge as early as 1952 of the damage to fetal lambs caused by radioactive particles. There was an "undisclosed Dr. Bustad report" that the AEC had but did not divulge in the first trial. (2) At least two staff medical experts, Lt. Colonel R.J. Veenstra and R.E. Thompsett, were pressured by AEC local staff to change their sheep autopsy reports due to the AEC's fears of lawsuits and adverse publicity. (3) The AEC's answers to the plaintiffs' 1956 interrogatories were evasive.

On the basis of the information produced at the May hearing,[15] Judge Christensen concluded that "improper means" had been used in the federal court, which were "unacceptable as part of the judicial process." Christensen concluded that these activities "clearly and convincingly demonstrate a species of fraud upon the court for which a remedy must be granted even at this late date."

These frauds defiled the court itself in that

> the judicial machinery cannot perform in the usual manner its important task of adjudicating cases that are presented for adjudication . . . By these convoluted actions and in related ways the processes of the court were manipulated to the improper and unacceptable advantage of the defendant at the trial . . . A court should not permit the very effectiveness of a party's improper tactics, nor even the lack of diligence of private parties, given a case of fraud on the court, to bar it from considering at any time the public interest and the imperative of safeguarding the judicial process and the court as an institution.

The federal judge harshly rebuked the AEC administrators for using improper means to achieve certain agency objectives: cover-up, deceit, and withholding information in discovery proceedings were "unacceptable as part of the judicial process." The federal court was "manipulated to the improper and unacceptable advantage" of the government. Judge Christensen found, in 1982, that

> the evidence *wrongfully withheld* at the trial, including the fetal lamb report, was of substantial significance and weight in favor of plaintiffs' claims for damages and was essential to a fair and proper consideration of the merits of those claims by the Court.[16]

(He rejected out-of-hand the government attorney's argument that the "missing report" would "have confused the Court.")

Documents were presented to the judge, and were part of the opinion[17] clearly indicating that the researchers assigned to the animal autopsies were pressured by AEC administrators into "abdicating long-time and forcefully defended positions that radiation was a contributing cause" in the animals' deaths. The AEC administrator involved in this coercion, Dr. Bernard F. Trum, wrote to the Assistant U.S. Attorney working on the case in 1956 that the doctors "will definitely state that they do not have evidence that radiation injury was either the cause or contributed to the death of the sheep," and that they would disqualify themselves.[18]

During the hearings in May, 1982, Dr. Trum testified that he made the decision in 1956 to introduce only "objective" evidence and that "to have injected into the trial contrary or circumstantial evidence [had] to be avoided to the extent reasonably possible." Judge Christensen concluded that these efforts were

> unjustified, improper, and incompatible with the judicial process, and yet to have been to a substantial degree effective in withholding from the Court or weakening evidence which the Court should have been permitted to consider on its own merits in connection with other evidence in this case.[19]

Finally, Judge Christensen focused on the government's attorneys' activities during the discovery phase:

> The answers to interrogatories prepared by defendent's attorneys were intentionally evasive and prejudicially misleading as heretofore found. The Court finds that at least one, if not more, of the defendant's lawyers was a party to an *intentional deception* in this respect [my italics].

Christensen reasoned that the judiciary must fashion a remedy even at this late date,[20] because of:

1. "false and deceptive representations of government conduct,"
2. "improper but successful attempts to pressure witness,"
3. "a vital report intentionally withheld,"
4. "information in another report presented in a deceitful manner,"
5. "interrogatories deceptively answered," and
6. "deliberate concealment of significant facts with reference to possible effects of radiation upon the plaintiff's sheep."[21]

A fraud was "perpetuated upon the court." The remedy for this action was harsh: vacating of the 1956 judgment and scheduling of new trials on the merits based on the evidence uncovered in 1979.

Why had the AEC administrators lied and deceived the federal judge in 1956? Christensen concluded that it was not for personal or monetary gain. The frauds were perpetrated for the "ultimate purpose [of] advancing the perceived interests of the United States in the unimpeded testing of nuclear weapons and to prevent this program from being needlessly interfered with or embarrassed by judicial action. But the improper means by which this purpose was sought to be achieved are, in my opinion, unacceptable as a part of the judicial process."[22]

Noting his 1956 deference to the integrity of the AEC scientists, Christensen, in 1982, wrote: "It is quite true that judged by modern insights I took a somewhat pristine view at the original trial of the general integrity of government officials in the absence of evidence impugning it in specific instances. I suppose that I shall continue to do so, in spite of the buffetings of Watergate, these proceedings, and other current disclosures."[23]

In a letter to the author, Judge Christensen wrote that he was moved to overturn his *own* 1956 opinion because of his "overriding commitment to law and legal processes not wholly unworthy of a judge."[24] His normative commitment "precluded in the first instance [1956] a decision based upon mere suspicion rather than evidence and, again [in 1982] led to the sustaining of that law and those processes when the evidence of their attempted subversion through fraud upon the Court became available."[25] The hope of this venerable judicial poet is that his *Bulloch* ruling will "strengthen the judicial system, . . . especially if the Supreme Court reaffirms its principle."[26]

On November 23, 1983, the U.S. Court of Appeals, Tenth Circuit, reversed and remanded the Christensen judgment in *David Bulloch et al. v. the United States*. The three-judge panel concluded that "the showing made by plaintiffs in the Bulloch II hearings falls far short of proof of fraud on the court or any other kind of fraud."[27] However, on April 2, 1984, the Tenth Circuit announced that the *Bulloch* case would be reviewed by the full court sitting *en banc*. In mid-May 1985, over 30 years after the original complaint was filed, the Tenth Circuit, sitting *en banc*, handed down its judgement in *Bulloch, II*. It overturned the Christensen decision. This decision will certainly be appealed to the U.S. Supreme Court by the plaintiffs.

Appendix B: The Medical Controversy over the Association Between Low Levels of Radiation and Cancer Occurrence

In examining this extended medical debate,[1] especially the question of nuclear fallout downwind of the Nevada Test Site, a number of specific questions have to be addressed. The extrapolation and proportionality dimension of medical research must be discussed, along with the consequences of extrapolation from known results of high doses of radiation exposure on human cohorts. In this regard, the problems involved in measuring doses of radioactivity need to be examined due to the number of serious questions raised about the dosage records maintained by the AEC and by the PHS during the period of above-ground testing. Finally, an examination of the nature of epidemiological investigations of low level radiation doses will be presented.

Due to the lack of low-level radiation exposure data available to medical researchers, intensive examinations must be performed on groups exposed to high levels of radiation. The researcher not only determines the impact of high doses on human cohorts but also speculates on the effects of low doses based on the high-dosage data. Probably the "most valuable human data available for the evaluation of late effects of radiation" come from studies of the Atomic Bomb Casualty Commission on the atom bomb survivors of Hiroshima and Nagasaki.[2]

All of our government's "risk estimates" derive from high-dose extrapolations such as the Hiroshima and Nagasaki studies.[3] We see the harmful effects of the high doses of radiation and then extrapolate the information collected since 1945 down to the low-dose range. To assist the researcher in this effort a model is needed to extend the curve to lower doses, "where judgments or educated guesses are usually required."[4]

In addition to the largely discredited threshold model, medical researchers exploring the questions of low-level radiation exposure and the relative risk and probability of cancer induction, use at least three risk estimate models (Figure 18), which are hypothetical and quantitative. Ironically, cancer researchers' theoretical postulating (in the absence of empirical knowledge) is largely the cause of the polarization within the scientific community.

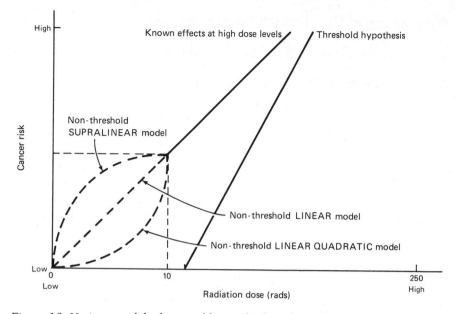

Figure 18. Various models discussed by medical epidemiologists when they discuss the associations between low levels of radiation and the onset of cancer.

"The disagreement concerns how to extrapolate from higher dose rates to the non-measurable range." The rancorous discord among scientists concerning the low-dose cancer danger is over hypotheses—not observable fact.[5]

The *Linear Hypothesis* Model, is based on the theoretical proposition that the damage caused per rad is *proportional*. "The number of radiation-induced cancers is directly proportional to the dose of radiation delivered, down to the lowest conceivable doses"; in short, the genetic and biological harm per rad does not vary between the high and low dose.[6] According to this hypothesis, the cancer risk at the lower dose levels is the same as the cancer risk at the known, recorded high levels. For example, if 10 out of 1,000 persons receiving 100 rads die of radiation-related deaths, there will be 10 out of 1,000,000 persons receiving 0.1 rad exposure who will die of radiation-related deaths.[7]

The *Linear-Quadratic* Model is based on the fairly new hypothesis that less damage occurs per rad at low levels of radiation exposure than at higher doses. Implicit in this hypothesis is the theoretical assertion—although there are no known empirical facts to buttress it—that the body has the ability to repair cells damaged by low-level radiation exposure.

The implicit question of the *cumulative* effects of low-dose exposure is therefore another issue that has caused controversy within the scientific community. "Tissues," wrote Gofman, "in most cases are able to regenerate after low-level radiation exposure. But above a certain radiation level, tissue does not retain the ability to regenerate."[8]

Therefore, some medical researchers have argued that it is inappropriate to apply the Linear Quadratic model to the downwinders since there were over 200 above-ground tests that produced a fairly *continuous* flow of low levels of radiation. For these researchers, the adverse health effects of the ionizing radiation are cumulative. Chronic exposure to the fallout radiation increases the dangers of cancer and leukemia because there is an "increase in the total amount of radiation delivered to a particular tissue."[9] For example, "chronic, repeated exposures of relatively small doses of radiation can increase the risk of leukemia severalfold."[10]

Users of the Linear-Quadratic model, however, theorize that the tissues continually repair themselves and that therefore less damage is caused per rad at low-dose levels than at high doses. The Quadratic model is more liberal than the Linear model: it is premised on the assumption that low doses will be over-estimated by a researcher linearly extrapolating from high doses.

The *Supralinear* Model proposes that *more* damage is caused per rad at low-dose levels than at the higher, known doses. Whereas high doses of radiation kill cells outright, as well as on some occasions, the human host, lower doses of radiation weaken and damage cells, which tend to live on in an altered, cancerous state. The model clearly "suggests the existence of large numbers of suprasensitive members in the population. A review of the epidemiology data has found no support for the existence of effects greater than those based on a linear model in humans."[11]

Of the three models used by medical researchers and government agencies to extrapolate data from high dose information, the linear model, the most conservative of the three, is the most widely used. All three models hypothesize, however, that there is *no threshold* that exists, below which exposure to radiation does not increase the risk of cancer. Even the federal agencies involved with nuclear research and atomic testing have come to accept the *no-threshold* premise. A 1980 Federal Interagency Report stated that "there may be no threshold dose level for most of the carcinogenic effects of radiation. Since the late 1950s, most scientists have accepted the no-threshold hypothesis, which holds that any dose of ionizing radiation can increase the risk of developing cancer in many organs and tissues of the body."[12]

However, these quantitative hypotheses are only educated guesses. Future discussions about qualitative epidemiological studies which focus on the Nevada Test Site and the downwind citizens will point out how these quantitative risk-estimate models do not account for individuality (age, biological variability, etc.) or environmental variability. Epidemiological studies which do take such factors into account can lead to more precise, accurate estimates of risk. Beyond this objective, epidemiological studies emphasize the distribution of a disease in space and time. They try, in the words of Dr. Joseph Lyon, to develop "an association between two events in time and in geography."[13]

Appendix C

- Excerpts from The Atomic Energy Commission's booklet, *Atomic Testing in Nevada,* 1957
- Brochure for Citizens Call
- The Downwinders' newsletter, "Testing News"

ATOMIC TESTS
IN
NEVADA

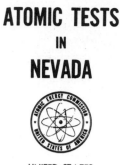

UNITED STATES
ATOMIC ENERGY COMMISSION
MARCH 1957

PROTECTION OF THE PUBLIC

You people who live near Nevada Test Site are in a very real sense active participants in the Nation's atomic test program. You have been close observers of tests which have contributed greatly to building the defenses of our country and of the free world. Nevada tests have helped us make great progress in a few years, and have been a vital factor in maintaining the peace of the world.

Some of you have been inconvenienced by our test operations. Nevertheless, you have accepted them without fuss and without alarm. Your cooperation has helped achieve an unusual record of safety. [...]

BULLETIN
ATOMIC TEST
SHOT SCHEDULED
FOR 5 AM
TOMORROW

Every Test Is Evaluated

Every test detonation in Nevada is carefully evaluated as to your safety before it is included in a schedule. Every phase of the operation is likewise studied from the safety viewpoint.

An advisory panel of experts in biology and medicine, blast, fallout, and meteorology is an integral part of the Nevada Test Organization. Before each nuclear detonation, a series of meetings is held at which this panel carefully weighs the question of firing with respect to assurance of your safety under the conditions then existing.

Warnings and Procedures

As in past series, every effort will be made to warn people away from the test site and the bombing range.

Helicopter and light aircraft sweeps of close-in predicted fallout areas will be made before a shot and any persons found there will be warned to leave. Like sweeps will be made following a shot. Stockmen will be advised if there are indications their stock has been exposed.

An extensive radiation monitoring system will be in operation in the test site region.

During the spring 1955 series, off-site monitoring was developed into an elaborate system to take numerous radiological measurements and also to provide close liaison with the residents of nearby communities. The U. S. Public Health Service stationed representatives in 12 zones east, northeast, and north of the site—the most frequent location of fallout. In addition, there were six mobile monitoring teams, staffed by the AEC and its contractors, on call to go to any locality if needed or to travel to areas outside the 12 zones. There were low-level aircraft sweeps to check ground fallout, as well as higher-level cloud tracking. Various automatic recording systems were used. Thousands of film badges were worn by selected individuals and placed on buildings throughout the area.

These procedures will be followed in the 1957 series. Two will be expanded: (1) More people will be asked to wear film badges, in the continuing effort to determine more precisely the exposures to individuals under varying conditions. (2) Public Health Service monitors will be stationed in new areas generally south of the test site, extending into Arizona and California, which were served previously by mobile teams.

If you are in an area exposed to fallout, you will be so advised by our radiation monitors who will explain just what is happening. If there is any probability that your exposure will approach our conservative guides, you will be advised what to do. For example, at St. George and Lincoln Mine, residents were advised to stay indoors for a few hours until the fallout had ceased. If you were outdoors during the fallout, you might be advised to bathe, wash your hair, dust your clothes, brush your shoes, etc. If the fallout pattern is across a highway, traffic might be halted temporarily.

Your best action is not be worried about fallout. If you are in a fallout area, you will be advised. If your radiation monitors advise precautionary action, do what they say. Please bear in mind that it is extremely un-

ROAD CLOSED

likely that there will be fallout on any occupied community greater than the past low levels. If you think that maybe you have been in fallout, or if you have other questions, get in touch with our monitors or with the Nevada Test Organization.

Outside of the test site region, there is also a need to obtain data on fallout for scientific purposes and to provide information to the public. A network of AEC monitors, of U. S. Weather Bureau Stations, and a U. S. Public Health Service network will monitor at localities across the Nation. [...]

THE FLASH OF LIGHT

The effects of the flash of light are essentially no different from those of sunlight. If you look directly into the sun (or at a photographer's flash bulb), you get black spots in front of your eyes for a few seconds or a few minutes. If you were much closer to the sun or if you used binoculars, eye damage might result.

On-site the thermal (heat) waves can injure eye tissues and cause permanent eye damage if one looks directly at the fireball. This is also true in the air above the test site. At shot time all personnel on or above the test site wear extremely dark glasses or turn away; binoculars are prohibited; and road traffic may be halted.

Off-site the same precautions should be followed by anyone in line of sight with the expected burst. The flash can cause "black spots" so that momentarily you can't see, or the flash can startle you if it is unexpected. This effect can be experienced at night many miles away. The greatest caution needs to be used by drivers of vehicles or the pilots of aircraft who might have an accident if momentarily unable to see, or if startled.

The brightness of the light striking your eyes depends of course on whether it is night or day (at night, more light enters the dilated pupils), whether there is direct line of sight to the fireball, on distance, on atmospheric conditions, and to some extent on the yield of the device. A majority of Nevada shots must be in the predawn hours of darkness and will require precautions against flash.

Past Experience With Flash

There have been no known cases of serious eye damage from light effects to people off-site. Some observers on nearby mountains, who did not wear dark glasses nor turn away, have reported temporary blind spots.

Off-Site Warnings and Procedures for Flash

Private and commercial air flights above the test site are prohibited. A circle about 65 miles in radius is established around the test site in which aircraft travel is restricted from 30 minutes before until 30 minutes after a shot (also because of radiation and air traffic congestion).

Some shots—because of time of day, very low yield,

or their positioning—do not require off-site precautions.

If precautions are indicated, the Nevada Test Organization will announce the scheduled time of the shot and will recommend precautions. These may include the following:

Day or predawn shots: Do not use binoculars or rifle scopes or other optical systems to look toward the test site at shot time. Do not look toward the test site at shot time unless you are wearing dark sunglasses.

Daytime shots: If the fireball will be visible on highways within a radius of up to 60 miles, a gen-

eral warning will be issued and insofar as possible those driving toward the test site will be warned of time of shot and advised to stop their cars and face away.

Predawn shots. If fireball will be visible on direct line to highways within a 60-mile radius drivers going toward the test site will be warned to stop at shot time. Persons in parked cars, or observers elsewhere, will be advised to look the other way or to wear two pairs of darkest variety sunglasses.

Radiation Is Nothing New

Very few of us can explain electricity, although we have learned to live with it and to use it. Even fewer can explain nuclear radiation. It is little understood by most of us, being something we can't see, feel, hear, taste, or smell.

And yet, radiation is nothing new. Since the beginning of time, mankind has been bombarded by radiation from outer space and from the ground beneath him. Cosmic rays rain down from space upon each of us every second of our lives. We are also constantly exposed to radiation from uranium, radium, and other elements in the earth itself. Each of us also has radioactive materials within his body. The sum total of this radiation is known as the background level.

Cosmic Rays Are Radiation

Cosmic rays at sea level give between 33 and 37 milliroentgens a year, depending on latitude and being least intense at the equator. At 5,000 ft. altitude, the dosages climb to between 40 and 60 mr. and at 15,000 ft. to between 160 and 240 mr.

The earth's surface everywhere is radioactive. Granite rock, for instance, contains radioactive radium, thorium, and potassium. Sea water contains little radiation. If you live on a ship on the ocean, exposure is greatly reduced.

Our bodies also contain radioactive materials, taken in with the food we eat and the water we drink. An AEC scientist illustrates his talks with an interesting experiment, in which the solid residue from a single sample of body fluid makes a geiger counter tick, as a piece of uranium ore would.

Your Body Contains Radiation

The natural radioactive carbon in your body exposes you to 1.5 mr. a year. The largest source of radioactivity in the body is potassium. It exposes you to 19 mr. a year. In fact, the radioactivity of the human body and the nature of its radiation are such that people receive radiation exposures from one another which are measurable. It has been calculated that people packed in a dense crowd would receive about 2 mr. a year dosage from the radioactive potassium in their neighbors' bodies.

217

CITIZENS CALL

A new awareness exists in this country—danger of nuclear attack and fear of its consequences. Those of us living downwind from the Nevada Test Site have been under nuclear fallout attack for thirty-one years. The price of this bitter legacy has been very high, and we are still reaping the consequences.

Nuclear Weapons Testing in Nevada

January 27, 1951, marked the dawn of the nuclear age in the Nevada desert. The nuclear weapons tests were conducted at the Nevada Test Site (NTS) until October, 1958. Each series of tests during that time was given a name such as Ranger, 1951; Upshot-Knothole, 1953; and Plumbob, 1957. The United States and Soviet Union observed a testing moratorium from October, 1958, to September, 1961, when the USSR launched a series of large tests. The United States quickly resumed its testing program.

Atmospheric testing included above-ground tests to develop and evaluate weapons systems and near-surface crater and shaft tests. Towers, balloons, airdrops, and cannons were the methods employed for the above-ground tests. The cratering program buried nucler devices in the desert rock. These explosives were tested for their use in non-military earthmoving activities.

Unfortunately for people living downwind (Utah, Nevada, and northern Arizona) from the NTS, radioactive fallout created by the atmospheric tests travelled far beyond the site's borders. Fallout was detected as far away as Canada and upstate New York. The tests were conducted when the prevailing winds were not blowing towards Las Vegas or California, the most-populated nearby areas. Off-site radioactive readings were recorded for approximately seventy percent of the atmospheric tests conducted at the NTS.

U.S. nuclear weapons tests are now conducted underground in accordance with the Limited Test Ban Treaty. Complete containment of the test is the goal of the program; however, radioactive fallout has vented or leaked from forty-three of the 470 NTS underground blasts. Twenty-eight of these vents are considered major—releasing about the same amount of radioactivity as the above-ground tests.

Human Effects of Nuclear Testing

The Forgotten Guinea Pigs, a Congressional report on the effects fallout has had on downwind victims, concludes that the federal government chose to protect its atmospheric weapons testing program rather than the health and welfare of area residents.

Recent studies show that the effects of low-level radiation may be seriously underestimated. Health researchers found a forty percent increase in the leukemia incidence for children born in Utah between 1951 and 1958 than the children born before or after that time. Children living in southern Utah—where fallout was heaviest—suffered an even higher leukemia incidence.

Three types of health damage are seen among the test site victims. These are cancer, including malignant blood diseases; genetic, somatic and teratogenic damage; and psychological stress.

Establishing proof that radiation exposures caused diseases in individuals is difficult. Cancer effects of low-level radiation can take up to thirty years or more to appear. Genetic damage can take several generations to appear. The difficulty is compounded by a pattern of withholding radiation and medical records, intimidating affected persons and their families, and suppressing pertinent scientific information.

Many fallout victims have expressed the feeling that they were used as guinea pigs. Most were never told about possible radiation hazards. None of these citizens were made aware of potential genetic and birth defects. They have not received any acknowledgement that there may be a radiation effect. Victims recall instances of arrogance and contempt by federal agencies, particularly by those in the Atomic Energy Commission and Department of Energy.

The personal tragedies of these individuals are the result of the conflict between promoting nuclear technology and protecting human health. The situation had led to a mistrust of the federal government: radiation victims feel they have no place to go for help. And the testing continues . . .

Why Citizens Call?

Citizens Call was founded in 1979 by radiation victims to fill a need not met by the government for citizens living in communities affected by NTS. Our efforts have been expanded to include anyone affected by nuclear weapons programs because the fallout from the NTS extends beyond Utah, Nevada, and Arizona. Radiation effects from uranium mining and milling, weapons fabrication, transportation, and storage of nuclear wastes are among our concerns.

Assisting victims is Citizens Call's primary purpose. We work actively to assure that what happened to us does not happen again. We cannot wait for others to act for us. Among Citizens Call's activities are:

- increasing public awareness and understanding of radiation issues including medical care, comprehensive nuclear test ban and compensation
- exploring the feasibility of building a cancer screening and care facility, and radiation center in southern Utah for fallout victims in the region
- starting a hospice program for cancer victims in areas downwind from NTS
- reconstructing the radioactive fallout doses received by radiation victims.

U.S. Nuclear Detonations[1]

Year	Number	Year	Number	Year	Number
1945	3	1961	10	1972	8
1946	2	1962	96	1973	9
1948	3	1963	43	1974	7
1951	16	1964	29	1975	16
1952	10	1965	28	1976	15
1953	11	1966	40	1977	12
1954	6	1967	28	1978	12
1955	18	1968	33	1979	14
1956	18	1969	29	1980	14
1957	32	1970	30	1981	17
1958	77	1971	12	1982	11[2]

[1] All announced nuclear tests conducted by the United States from July, 1945 through August, 1982 are listed above. Until July, 1962, atmospheric and underground nuclear tests were conducted at the NTS, and atmospheric and underwater tests were conducted in the Marshall Islands, Christmas Islands, and Johnston Island areas of the Pacific, and over the Atlantic Ocean. Most underground tests since July, 1962, have been conducted at the NTS. Some tests were conducted on the Nellis Air Force Base Bombing Range; on Amchitka, one of the Aleutian Islands off the coast of Alaska; in central and northwestern Nevada; and in Colorado, New Mexico and Mississippi.

[2] As of August 5, 1982.

Joining Citizens Call

Citizens Call was founded by downwind victims and concerned family members, convinced that we must help ourselves to learn the facts, share them with others, and find a way to organize ourselves and help people who are still suffering. We must also stop the current underground nuclear weapons testing and prevent this kind of public policy from ever plaguing another group of Americans.

Your Citizens Call membership includes

- a subscription to *Citizens Voice*, Citizen Call's quarterly newsletter
- voting privileges at Citizens Call Annual Meeting held in the Fall
- special updates and bulletins on radiation issues.

218

TESTING NEWS

VOL. III No. 5
January 27, 1985

The only thing standing between you and the Nevada Test Site

ARMS TALKS UNDERWAY SOON
A QUESTION OF SINCERITY

By Monte Bright

Since the Strategic Arms limitation Talks (START) broke up in November 1983 the relations between the two super powers have been extremely cold. The Soviets had made it known that they would terminate all strategic arms discussions if the United States went ahead with its deployment of Euro-missiles. The United States chose to begin deployment and the Soviets, true to their word broke off discussions.

BOTH SIDES LOSE FACE

For more than a year the Soviets have repeatedly said they would not resume negotiations until the United States removed the newly added Pershing and Cruise missiles from Europe. The United States refused and hopes of any break in resuming talks were deadlocked. At the same time the Reagan Administration was mouthing its desire to "rid the world of all nuclear weapons", in an attempt to convince voters that they were sincere in their expressed desire to reach an arms control agreement with the Soviets. While both sides publicly declared a desire to limit the growing arms race, both eventually resorted to little more than exchanging insults and the arms race continued unabated. Many experts claimed The Soviets had gambled they could ride out four years of the Reagan Administration in the hope that Reagan would lose the next election. Some said that the Soviet refusal to continue arms talks was nothing more than rhetoric designed to hurt Reagan at the polls in November, and others claimed that the Reagan Administration had made it impossible for the Soviets to negotiate and had failed to provide a firm platform for any meaningful discussions to take place. But as the election neared and it was apparent that Reagan would have a land slide victory they made a surprising change of face and sent Soviet Foreign Minister Gromyko to Washington to discuss the matter directly with Reagan. Administration supporters said that the Soviets had realized that they had failed to affect the election and now had to eat their words, being isolated by world opinion.

Once the election had passed and both sides were faced with four more years of the status quo plans moved forward to resume talks. United States Secretary of State George Shultz met in Geneva with Soviet Foreign Minister Andrei Gromyko in early January and both sides came to an agreement resulting in an announcement just last week that talks would resume in Geneva on March 12th.

The real substance of the upcoming talks still remain hazy and other than broad generalities it isn't fully known what will be discussed. The Soviets seemed concerned with reaching an agreement banning space weapons, while the United States has refused to stop research and development for the Strategic Defense Initiative (SDI), or "Star Wars", maintaining that it is the only plausible way to ever halt the arms race. [...]

[...] The provisions required to enact a ban on nuclear testing have been agreed on by the Soviet Union, the United States and Great Britain and all that is required is the political will on the part of the administration.

The Reagan Administration was the one who broke off negotiations for no reason on a comprehensive test ban when an agreement was in sight. They could easily resume test ban talks and have a verifiable agreement that would definitely provide the first step toward halting and reversing the arms race. If sincerity was not the underlying problem of Reagan's approach to arms control by ending testing we could acheive an end to the qualitative component of the arms race.

The Soviet Union has stated, and continues to state its willingness to immediately ban testing. If WE really want arms control and if the problem of sincerity is really THEIR´S, and not partly ours too, then the Reagan Administration should announce an immediate moratorium on nuclear testing and call for talks to conclude a test ban. If the Soviets are not sincere then so be it, if on the other hand they are on this point, then end testing and begin the long process needed to rid the world of nuclear weapons.

Talks about talks that talk about doing something is not the same as sincerely addressing the issue. More and more weapons development peace does not make. As President Kennedy said during the debate over atmospheric testing, "We test and then they test, and we have to test again...and you build up until somebody uses them."

****

NO MORE LIES
WESTERN TOUR

Starting on February 1st in Los Angeles and concluding in Reno in early March an eight state organizing and information sharing tour will focus attention on the nuclear testing issue. The Downwinders´ sponsored tour will help to spread educational information and grass roots organizing skills to areas of the western United States concerned with issues related to nuclear testing and nuclear weapons facilities.

The tour, entitled, NO MORE LIES, will "bring new information on developments in nuclear weapons facilities, the cumulative effects of radiation exposure from the U. S. nuclear testing program, and the crisis in Central America. It will point out the long record of lies on the part of the federal government in promoting a continued arms race. Much of the tour will focus on the coverup of health hazards regarding the nuclear testing program at the Nevada Test Site. A letter sent out by tour organizers discusses a primary theme of the tour. "For the first time we have been able to compile enough information about the fallout from both aboveground and underground testing to better under-stand the impact of testing on our health." As part of the tour a full size poster has been developed show-ing the extent of fallout spread and exposures that resulted across the U.S. from testing at the Nevada Test Site. The poster, titled "The Bombing of America" is 19" x 25" and graphically demonstrates that not just the residents of Utah- Nevada and northern Arizona, but most Americans are, by virtue of the Nevada Test Site, The poster is two colors, on 100lb glossy paper, and is available for $3.00 from Downwinders.

The tour has two experienced and well versed speakers, Ken Nightingale, former staff member of the Livermore Action Group, researcher and writer on nuclear weapons and testing issues, and currently an organizer for Lab Watch, and Ada Sanchez, from Portland former national coordinator for the Supporters of Silkwood, organizer, speaker and writer on militarism issues. They are both active in grass roots work in their areas and have many organizing skills to share with local groups and organizers.

The tour has an impressive itinerary; Los Angeles February 2, San Diego February 4, Tucson, Arizona February 6-7, Mesa, February 8, Flagstaff February 9-10 and into New Mexico with stops in Albq on the 12-13, and Santa Fe on the 14th. The tour will resume in Boulder, Colorado on February 26, move into Wyoming for stops in Burns and Laramine the 27th and 28th. The tour will be in Salt Lake City on March 1, and spend four days in Idaho with stops in Boise on March 2, Twin Falls March 3, Pocatello March 4, and Ketchum on March 5. The tour will finish up with work in the Reno area.

For additional information on the tour in your area contact Downwinders, 1321 East 400 South, Salt Lake City, Utah, 84102, (801) 583-5252.

A full report on the tour will run in the March issue of Testing News.

THANK GOD FOR CIVIL DEFENSE

by Ken Nightingale

Recently, the Federal Emergency Management Agency (FEMA), the agency in charge of devising ways to help the American public survive nuclear attacks, rejected a survival study prepared by the Lawrence Livermore Lab oratory as "ludicrous" and "absurd". Robert Hickman, the author of the study, advised workers to submerge themselves four feet deep in a body of water for up to 30 minutes as a shield against the immediate effects of a nuclear blast. FEMA cancelled its $174,000 contract with Livermore and refused to pay for the study.

"A body of water could provide a unique protective option for some individuals. Considerable protection could be obtained from the prompt nuclear effects by wearing as much clothing as possible, diving about four feet down and spending as little time as possible at the surface for air."

"However, workers taking advantage of large bodies of water should not only be good swimmers but they should also tether themselves to a flotation device with a 10-foot line."

In a August 29 memorandum, Richard Feirman, FEMA's project manager, complained "the report is ludicrous;" "conclusions and recommendations are absurd."

"Unfortunately, one is of the opinion that the LLNL author is making light of a most serious subject, namely survival."

The San Francisco Chronicle carried this story on October 22. When the author was given a chance to answer the charges in a story the next day he stood by his proposal: "If I thought it was unreasonable, I wouldn't have put it into the report," he said.

Who should be embarrassed, FEMA or the Livermore Lab? Both. The lab should be embarrassed to propose such an impractical solution to the serious problem of surviving a nuclear attack. Imagine the impact of posters in every factory across the country advising workers to jump in the lake in the event of a nuclear attack. But why should FEMA be embarrassed? In their righteous criticism of the Lab they do not realize that the joke is on them. [...]

221

Notes

CHAPTER ONE

1. Henry DeWolf Smyth, *Atomic Energy for Military Purposes: The Official Report on the Development of the Atom Bomb under the Auspices of the U.S. Government, 1940–1945* (Princeton, New Jersey: Princeton University Press, 1945), pp. 1–2.

2. *Ibid.*, p. 22.

3. James W. Kunetka, *City of Fire: Los Alamos and the Atomic Age, 1943–1945,* (Albuquerque: University of New Mexico Press, 1979), pp. 74–77.

4. Alan P. Lightman, "To Cleave an Atom," *Science,* 5 (November, 1984), p. 105; Kunetka, *City of Fire,* p. 17.

5. Leslie M. Groves, *Now It Can Be Told: The Story of the Manhattan Project* (New York: Da Capo Press, 1962), p. 5.

6 Lightman, "To Cleave an Atom," p. 105; Daniel J. Boorstein, *The Americans: The Democratic Experience* (New York: Random House, 1973), p. 583, Smyth, *Atomic Energy,* p. 24.

7. Kunetka, *City of Fire,* p 18

8. Lightman, "To Cleave an Atom," p. 105.

9 *Ibid*

10. Richard G. Hewlett and Oscar E. Anderson, Jr., *The New World, 1939–1946: A History of the U.S. Atomic Energy Commission* (University Park, Pennsylvania: University of Pennsylvania Press, 1962), p. 10.

11. Boorstein, *The Americans,* p. 582.

12. Smyth, *Atomic Energy,* p. 45.

13. Groves, *Now It Can Be Told,* p. 6.

14. *Ibid.*, p. 5.

15. Quoted in Morton Grodzins and Eugene Rabinowich, *The Atomic Age: Scientists in National and World Affairs* (New York: Basic Books, 1963), pp. 11–12.

16. Hewlett and Anderson, *New World,* p. 17.

17. Kunetka, *City of Fire,* p. 6.

18. Groves, *Now It Can Be Told,* p. 38.

19. Senate Special Committee on Atomic Energy, *Atomic Energy,* Hearings, pt. 1, 79th Congress, 1st session, November 27–30, 1945, p. 335.

20. *Ibid.*, p. 37.

21. "Victor Weisskopf, Physicist," on *Nova,* Public Broadcasting Television Network, September, 1984.

22. *Ibid.*

23. Kunetka, *City of Fire,* p. 73.

24. Groves, *Now It Can Be Told,* p. 265.

25. Kunetka, *City of Fire,* p. 14.

26. Hewlett and Anderson, *New World,* p. 3.

27. Smyth, *Atomic Energy,* p. 85.

28. See generally, Groves, *Now It Can Be Told.*

29. Smyth, *Atomic Energy,* pp. 85–88.

30. Tyler Abell, ed., *Drew Pearson, Diaries, 1949–1959* (New York: Holt, Rinehart and Winston, 1974), pp. 70–71.

31. Groves, *Now It Can Be Told*, p. 40.
32. *Ibid.*, p. 149.
33. Kunetka, *City of Fire*, p. 34.
34. Smyth, *Atomic Energy*, pp. 207–209.
35. "Weisskopf," *Nova*.
36. Groves, *Now It Can Be Told*, p. 88.
37. Kunetka, *City of Fire*, p. 60.
38. Hewlett and Anderson, *New World*, p. 250.
39. Kunetka, *City of Fire*, p. 75.
40. Smyth, *Atomic Energy*, p. 253.
41. *Ibid.*, p. 247.
42. Kunetka, *City of Fire*, pp. 168–169.
43. Hewlett and Anderson, *New World*, p. 379.
44. "Weisskopf," *Nova*.
45. Smyth, *Atomic Energy*, p. 253.
46. "Weisskopf," *Nova*.
47. William Lawrence, The *New York Times*, July 18, 1945.
48. Quoted on KUTV, "Downwind," Salt Lake City, Utah, December 17, 1982.
49. For an extensive study of the devastation of Hiroshima and Nagasaki, see generally, Committee for Compilation of Materials on Damage Caused by the Atom Bombs in Hiroshima and Nagasaki, trans. Eisei Ishikawa and David L. Swain, *Hiroshima and Nagasaki: The Physical, Medical, and Social Effects of the Atomic Bombings* (New York: Basic Books, 1979, 1981).
50. *Ibid.*, pp. 30, 67. The authors clearly point out the difference between an atomic bomb and a conventional bomb:

> The power of the atomic bomb as usually expressed in equivalents of TNT is merely a comparison of total amounts of energy and ignores the essential qualitative difference between an atomic bomb and a conventional bomb. The content of the energy yield by nuclear fission of Uranium 235 is of special importance. The emitted fission fragments have marked radioactivity, and about 17 percent of the total energy released by nuclear fission is the result of radiation. It is, therefore, inadequate to evaluate the power of the atom bomb in terms of the energy released by TNT. . . . An enormous amount of radioactivity, initial and residual, was released by the atom bombs over Hiroshima and Nagasaki—a crucial difference from the conventional bomb.

51. *Ibid.*, p. 32.
52. See, for example, Smyth, *Atomic Energy*, pp. 41–43.
53. Groves, *Now It Can Be Told*, p. 291.
54. *Atomic Energy*, 1945 Hearings, p. 37.
55. Reporter's transcript, Eighth Dose Assessment Advisory Group, U.S. Department of Energy, October 20, 1983, Nevada Operations Office, p. 274.
56. House Subcommittee on Oversight and Investigations, Committee on Interstate and Foreign Commerce, *The Forgotten Guinea Pigs: A Report on Health Effects of Low Level Radiation Sustained as a Result of the Nuclear Weapons Testing Program Conducted by the United States Government*, 96th Congress, 2d session, August, 1980, p. 39.
57. *Ibid.*, pp. 39–40.
58. Gregg Herkin, *The Winning Weapon: The Atomic Bomb in the Cold War, 1945–1950* (New York: Knopf, 1980), pp. 3; 20; 21.
59. *Ibid.*, p. 197.
60. *Ibid.*, pp. 98; 112; 125–126.
61. *Atomic Energy*, 1945 Hearings, p. 62.
62. Herkin, *The Winning Weapon*, p. 153.
63. Robert J. Donovan, *The Tumultuous Years: The Presidency of Harry S. Truman, 1949–1953* (New York: Norton, 1967), 1982, p. 30.

64. *Atomic Energy,* 1945 Hearings, pp. 234; 348; 130.

65. *Ibid.,* p. 186.

66. *Ibid.,* p. 32.

67. "Weisskopf," *Nova.*

68. Grodzins and Rabinowich, *Atomic Age,* pp. 19–21; *Atomic Energy,* 1945 Hearings, p. 304. Dr. John Simpson, a physicist who worked at Los Alamos Scientific Laboratory said that "after Hiroshima and Nagasaki there arose a rising tide of thought among scientists which showed an astonishing unanimity of opinion. Men who were isolationist in spirit, men who were convinced that war had a place in society, men of every shade of political thought, were all brought together" to battle against nuclear proliferation in the post-war world.

69. Quoted in *Progressive* (November, 1979), p. 17.

70. *Ibid.*

71. John G. Clark, David M. Katzman, Richard D. McKinzie, and Theodore A. Wilson, *Three Generations in Twentieth Century America: Family Community and Nation* (Homewood, Illinois: Dorsey Press, 1977), p. 399.

72. Herken, *The Winning Weapon,* p. 112.

73. Hewlett and Anderson, *New World,* pp. 1–2.

74. Herken, *The Winning Weapon,* p. 242; *Atomic Energy,* 1945 Hearings, pp. 31–32.

75. For an extensive account of the responsibilities and difficulties the AEC had in the late 1940s after the Cold War began in earnest, see Richard G. Hewlett and Francis Duncan, *Atomic Shield, 1947–1952, Volume II, A History of the U.S. Atomic Energy Commission* (University Park: Pennsylvania State University Press, 1969).

CHAPTER TWO

1. H. Peter Metzger, *The Atomic Establishment* (New York: Simon and Schuster, 1972), p. 18.

2. Joint Committee on Atomic Energy, *Atomic Energy Legislation Through the 94th Congress, 2d session,* 95th Congress, 1st session, March, 1977. Appendix C, Index to the Legislative History of the Atomic Energy Act of 1946, pp. 376–382.

3. Hewlett and Duncan, *Atomic Shield,* p. 170.

4. Hewlett and Anderson, *New World,* p. 632.

5. *Ibid.,* p. 434.

6. See *Atomic Energy,* 1945 Hearings.

7. Hewlett and Anderson, *New World,* pp. 429ff.

8. *Atomic Energy Act of 1946,* Public Law 79-85, 79th Congress, 1st session, August, 1946.

9. See *Atomic Energy Act of 1946* and Richard O. Niehoff, "Organization and Administration of the U.S. Atomic Energy Commission," *Public Administration Review,* 8 (Spring, 1948), p. 93. The legislative committee, responsible for AEC oversight, was "an unusual vehicle of coordination between the operating policy developed by the AEC and legislative policy. . . . It was clearly an outgrowth of the concern that the ordinary congressional controls secured through voting, appropriations, reporting and similar services were not enough."

10. *Atomic Energy Act of 1946.*

11. Niehoff, "Organization and Administration of the AEC," p. 94.

12. Hewlett and Duncan, *Atomic Shield,* p. 56.

13. *Ibid.,* p. xiv.

14. See, for example, Hewlett and Anderson, *New World,* pp. 625ff., and Groves, *Now It Can Be Told.*

15. Hewlett and Anderson, *New World,* p. 626.

16. *Ibid.,* p. 630.

17. Joint Committee on Atomic Energy, *Atomic Energy,* Hearings, 81st Congress, 1st session, February, 1949, p. 4. Other problems enumerated by Gordon Dean included: better budget and accounting reporting, cooperation and exchange of information with industrial firms not actually participating in the program, personnel security clearance procedures, and general improvement of the AEC's policies and procedures with regard to the issuing and controlling of information.

18. Joseph C. Goulden, *The Best Years, 1945–1950* (New York: Atheneum, 1976), p. 262.

19. Hewlett and Duncan, *Atomic Shield,* p. 360.

20. Niehoff, "Organization and Administration of the AEC," p. 95.

21. Hewlett and Duncan, *Atomic Shield,* p. 316; Niehoff, "Organization and Administration of the AEC," pp. 97–98; Hewlett and Anderson, *New World,* p. 640. There were at this time other national laboratories at Berkeley, California; Ames, Iowa; and Rochester, New York. *Atomic Energy,* 1949 Hearings, p. 13.

22. Niehoff, "Organization and Administration of the AEC," p. 95.

23. *Ibid.*

24. Hewlett and Duncan, *Atomic Shield,* p. 316. The AEC eventually evolved as a "skeletal bureaucracy" in Washington, D.C. With only a small number of federal employees, the AEC's contract with the private sector and with universities absorbed a large part of its annual budget. The following tabulation of AEC appropriations, from 1947 to 1950, shows general expenditures and the annual growth of the contract segment of AEC budgets:

AEC Appropriations, By Fiscal Year, 1947–1950

Appropriation	Area	Amount Funded (in millions)
1947	(MED funds transferred in January)	$608.5
1948	Cash	175
	Contract Authority	400
1949	Cash	511.8
	Contract Authority	650
1950	Cash	703
	Contract Authority	841

Atomic Energy Budget, in *Atomic Energy Legislation,* 1977 Report, p. 440.

25. Hewlett and Duncan, *Atomic Shield,* p. 360.

26. *Ibid.*

27. Goulden, *The Best Years,* p. 277. Seventy American military personnel, mostly airmen, lost their lives during this year-long airlift.

28. Donovan, *Tumultuous Years,* p. 102.

29. Joint Committee on Atomic Energy, *Atomic Energy Legislation,* 1977 Report, p. 440.

30. Hewlett and Duncan, *Atomic Shield,* pp. 381–382.

31. Donovan, *Tumultuous Years,* p. 150.

32. Metzger, *Atomic Establishment,* p. 72.

33. Hewlett and Duncan, *Atomic Shield,* p. 387.

34. Donovan, *Tumultuous Years,* p. 104.

35. America detonated its first hydrogen bomb in 1952; the Russians detonated their super bomb in 1953.

36. Joint Committee on Atomic Energy, *Soviet Atomic Espionage,* 82nd Congress, 1st session, April 1951, p. ii.

37. *Ibid.,* p. 6.

38. *Ibid.*, p. 5.

39. Department of Defense, *Operation Buster-Jangle, 1951, U.S. Atmospheric Nuclear Weapons Test, Personnel Review,* DNA 6-23F, 1982, pp. 19–20.

40. *Nevada Test Site,* Energy Research and Development Administration, *Final Environmental Impact Statement, Nye County, Nevada,* 1977, ERDA Rept. 1551, pp. 2–12.

41. Hewlett and Duncan, *Atomic Shield,* p. 535.

42. Department of Defense, *Operation Ranger, 1951, U.S. Atmospheric Nuclear Weapons Test, Personnel Review,* DNA 6-22F, 1982, p. 18. There were sensitive public relations and political reasons for the AEC not testing on the continental United States until there was an obvious necessity for the testing to take place there. However, the outbreak of the Korean conflict in June, 1950, resulted in a shift of the Commission's efforts from peaceful to military pursuits. With the start of hostilities, the AEC commissioners saw a need for a local testing site that would turn over military test data on new atomic devices to the scientists quickly so that the weapons loop process would not be delayed. See Hewlett and Duncan, *Atomic Shield,* p. 485ff.

43. House Subcommittee on Oversight and Investigations, Committee on Interstate and Foreign Commerce, *The Forgotten Guinea Pigs: A Report on Health Effects of Low Level Radiation Sustained as a Result of the Nuclear Weapons Testing Program Conducted by the United States Government,* 96th Congress, 2nd session, August, 1980, p. 24.

44. AEC Minutes, Meeting No. 504, December 12, 1950.

45. *Ibid.*

46. *Ibid.*

47. *Ibid.*

48. In 1974 Congress reorganized the Atomic Energy Commission. The Nuclear Regulatory Commission (NRC) took over the AEC's licensing and operating of nuclear power plants and the Energy Research and Development Administration (ERDA) assumed the AEC's research and development activities. In 1977, Congress transferred ERDA's research and development authority to the U.S. Department of Energy, which now produces atomic weapons systems for the Department of Defense.

49. Letter, Military Liaison Committee to Gordon Dean, Chairman of the AEC, July 16, 1951, quoted in Allan Favish, "Radiation Injury and the Atomic Veteran: Shifting the Burden of Proof on Factual Causation," *Hastings Law Journal,* 32 (March, 1981), p. 934, n. 6.

50. *Ibid.*

51. Department of Defense, *Operation Upshot-Knothole, 1953, U.S. Atmospheric Nuclear Weapons Tests, Personnel Review,* DNA #6014F, 1982, p. 31.

52. Favish, "Radiation Injury and the Atomic Vet," p. 944, n. 39.

53. *Ibid.*, p. 945.

54. *Ibid.*, p. 946. At a February 25, 1953 AEC Meeting, No. 825, AEC Commissioner Smyth responded to a Department of Defense request that news correspondents be allowed into the forward test area at the Nevada Test Site by stating: "If the DOD wished to accept full responsibility for these correspondents, the AEC would have no objection."

55. Department of Defense, *Operation Tumbler-Snapper, 1952, U.S. Atmospheric Nuclear Weapons Tests, Personnel Review,* DNA #6019F, 1982, p. 7.

56. Energy Research and Development Administration, *Final Environmental Impact Statement, Nevada Test Site, Nye County, Nevada,* September 1977, ERDA Rept. 1551, pp. 2–3.

57. Testimony of Mahlon E. Gates, Manager, Nevada Operations Office, U.S. Department of Energy, Before the House Subcommittee on Oversight and Investigation, Committee on Interstate and Foreign Commerce, April 23, 1979.

58. *Ibid.*

59. Department of Defense, *Operation Buster-Jangle, 1951, U.S. Atmospheric Nuclear Weapons Test, Personnel Review,* DNA-6-23F, 1982, p. 30.

60. *Ibid.*

61. Joint Committee on Atomic Energy, *The Nature of Radioactive Fallout and Its Effects on Man,* Hearings, 85th Congress, 1st session, May–June, 1957, p. 179.

62. AEC Minutes, Meeting No. 865, May 21, 1953.

63. *Ibid.*

64. *Ibid.*

65. See, generally, Joel D. Aberbach and Bert A. Rockman, "Clashing Beliefs within the Executive Branch," *American Political Science Review,* 70 (June, 1976).

66. AEC Minutes, Meeting No. 863, May 18, 1953.

67. AEC Minutes, Meeting No. 1,032, October 5, 1954.

68. AEC Minutes, Meeting No. 504, December 13, 1950.

69. Michael Uhl and Tod Ensign, *GI Guinea Pigs* (Chicago, Westview, 1980), p. 24.

70. Joint Committee on Atomic Energy, *Fallout from Nuclear Weapons Tests,* Summary of Hearings, 86th Congress, 1st session, August, 1959, p. 8.

71. "The Occurrence of Malignancy in Radioactive Persons," *American Journal of Cancer,* 15 (1931), p. 2435; Furth and Furth, "Neoplastic Diseases Produced in Mice by General Irradiation with X-Rays," *American Journal of Cancer,* 28 (1936), p. 54; Ulrich, "The Incidence of Leukemia in Radiologists," *New England Journal of Medicine,* 234 (1916), p. 45; Brues, Finkel, and Lisco, "Carcinogenic Properties of Radioactive Fission Products and of Plutonium," *Radiology,* 49 (1947), p. 361; March, "Leukemia in Radiologists in a 20-Year Period," *American Journal of Medical Science,* 220 (1950), p. 282; Furth, "Recent Studies on the Etiology and Nature of Leukemia," *Blood,* 6 (1951), p. 964; Folley, Borges, and Yamawaki, "Incidence of Leukemia in Survivors of the A-Bomb in Hiroshima and Nagasaki," *American Journal of Medicine,* 13 (1952), p. 311; Lange, Maloney, and Yamawaki, "Leukemia in A-Bomb Survivors," *Blood,* 9 (1954), p. 574; Lewis, "Leukemia and Ionizing Radiation," *Science,* 125 (1957), p. 956; "The Biological Effects of Atomic Radiation," National Archives of Science, National Research Council, 1956; *The Hazards to Man of Nuclear and Allied Radiations,* British Medical Research Council, 1956; See also *Radiation and Health,* bibliography, in *Radioactive Fallout,* 1957 Hearings, pp. 1881ff.

72. Favish, "Radiation Injury and the Atomic Vet," p. 938.

73. *Radioactive Fallout,* 1957 Hearings, p. 14.

74. *Ibid.,* p. 1200.

75. Hewlett and Duncan, *Atomic Shield,* p. 130.

76. Metzger, *Atomic Establishment,* p. 88.

77. Leslie J. Freeman, *Nuclear Witnesses* (New York: Norton, 1982). pp. 168–169.

78. *Radioactive Fallout,* 1957 Hearings, pp. 1202ff.

79. Subcommittee on Research, Development and Radiation, Joint Committee on Atomic Energy, *Radiation Standards, Including Fallout,* Hearings, 87th Congress, 2d session, June 4–7, 1962, pp. 26, 28.

80. House Subcommittee on Oversight and Investigations, Committee on Interstate and Foreign Commerce, and the Senate Scientific Research Subcommittee of the Labor and Human Resources Committee of the Committee on the Judiciary, *Health Effects of Low Level Radiation,* Joint Hearings, 96th Congress, 1st session, April, 1979, pp. 204–205.

81. *Radioactive Fallout,* 1957 Hearings, p. 20.

82. *Ibid.,* p. 21.

83. *Ibid.,* p. 103.

84. *Fallout from Nuclear Weapons Tests,* 1959 Hearings, p. 16.

85. *Ibid.,* p. 6.

86. *Radiation Standards,* 1962 Hearings, p. 6.

87. AEC Meeting, No. 862, May 13, 1953.

88. *Radioactive Fallout,* 1957 Hearings, p. 16. In the 1959 Joint Committee Hearings on Radiation, Dr. Dunham, of the AEC's Division of Biology and Medicine struck a slightly different tone about the risks associated with radioactive fallout; "Those immediately concerned with weapons development have made it one of their prime objectives to develop methods for reducing fallout. . . . *If we believe that the reasons for accepting any particular risk are sufficiently compelling, we accept them; if not, we choose an alternate course of action which avoids the risk. In doing so, we may either sacrifice a desired goal or substitute a more acceptable risk." Fallout from Nuclear Weapons Tests,* 1959 Hearings, pp. 10, 42.

89. Joint Committee on Atomic Energy, *Health and Safety Problems and Weather Effects Associated with Atomic Explosions,* Hearings, 84th Congress, 1st session, April 15, 1955, p. 7.

90. *Radioactive Fallout,* 1957 Hearings, pp. 1221–1222.

91. Plaintiffs' pretrial statement, *Allen v. United states,* 1982, pp. 4–5. In April, 1957, the AEC Biology and Medicine staff developed safety standards, approved by the AEC commissioners, for the operatives at the NTS to employ for themselves and for the downwinders. If the number of rads reached 10, the AEC personnel in the field were instructed to (1) make sure that the people downwind "remained indoors," (2) and if outdoors [for them] to be advised to change clothing and bathe, with the clothing cleaned by normal means," (3) decontaminate motor vehicles, and (4) take samples of water, air, food stuff, etc. See Gordon Dunning's testimony at the 1957 Hearings, *Radioactive Fallout,* p. 179ff.

92. *Health Effects of Low Level Radiation,* 1979 Hearings, pp. 2279–2280.

93. *Ibid.,* pp. 2281–2282.

94. *Ibid.,* p. 2273.

95. Plaintiffs' pretrial statement, *Allen v. United States,* 1982, pp. 42–55.

96. Frank Butrico, quoted on KUTV, "Downwind," Salt Lake City, December 17, 1982.

97. *Ibid.*

98. *Ibid.*

99. *Health Effects of Low Level Radiation,* 1979 Hearings, p. 258.

100. *Ibid.,* p. 259. Dr. Michael May, Department of Energy's Lawrence Livermore Radiation Laboratory, Berkeley, California, wrote that "estimates of dosage delivered by radioactive iodine to the thyroids of children in St. George, Utah who were less than five years old in 1953, vary between 500 and 2,500 rads." See Carl J. Johnson, "Cancer Incidence in an Area of Radioactive Fallout Downwind from the Nevada Test Site," *Journal of the American Medical Association,* 251 (January 13, 1984). p. 231.

101. *Health Effects of Low Level Radiation,* 1979 Hearings, p. 2130.

102. *Ibid.*

103. Harold Knapp, "Observed Relations Between the Deposition Level of Fresh Fission Products from Nevada Tests and the Resulting Levels of I-131 in Fresh Milk," AEC Report, March, 1963.

104. Dr. Edward S. Weiss, "Leukemia Mortality Studies in Southwest Utah," AEC Report, 1965.

105. John Gofman et al., "Low Dose Radioactivity, Chromosomes and Cancer," IEEE, October, 1969.

106. Freeman, *Nuclear Witnesses,* p. 90.

107. Uhl and Ensign, *GI Guinea Pigs,* p. 15.

108. Thomas H. Saffer and Orville E. Kelly, *Countdown Zero* (New York: Penguin, 1982), p. 263.

109. A. C. Titus, "Government Responsibility for Victims of Atomic Testing: A Chronicle of the Politics of Compensation," *Journal of Health Politics, Policy and Law,* 8 (Summer, 1983)., pp. 280–281.

110. *Ibid.,* p. 281.

111. Quoted on KUTV, "Downwind," December 17, 1983.

CHAPTER THREE

1. See Goulden, *The Best Years,* pp. 3–18, and Metzger, *The Atomic Establishment,* p. 53.

2. The Nevada Test Site (until 1954 referred to as the Nevada Proving Grounds) was selected by the AEC from among a final group of three sites (the others were Dugway, Utah and Alamogordo, New Mexico) for use as a "supplemental," "emergency alternative" to Eniwetok Island in the Pacific. While no site "can be considered a completely satisfactory alternative to overseas site," the NTS was the most favored continental site. It was selected because of its radiological safety figures (low population density and favorable meteorological conditions, and good physical conditions and availability of land.) "Because it was virtually uninhabited, had no substantial population within nearly 70 miles, was next to an existing Air Force gunnery range, afforded maximum security and *the prevailing winds blew east over southern Utah,*" the AEC chose the NTS. In 1950, however, the AEC conditioned the use of atomic bomb tests 'involving relatively low orders of energy release." *Health Effects of Low Level Radiation,* 1979 Hearings, Vol. II, pp. 1412, 1413, 1418, and Saffer and Kelly, *Countdown Zero,* p. 31.

3. Marvin S. Hill, "The Rise of the Mormon Kingdom of God," in Richard Poll, General Ed., *Utah's History* (Provo, Utah: Brigham Young University Press, 1978), p. 97.

4. *Ibid.,* p. 98.

5. Robert Gottlieb and Peter Wiley, *America's Saints: The Rise of Mormon Power* (New York: G. P. Putnam's Sons, 1984). p. 32.

6. Hill, "Mormon Kingdom of God," p. 98.

7. *Ibid.*

8. See, generally, James B. Allen and Glen M. Leonard, *The Story of the Latter Day Saints* (Salt Lake: Deseret Book Co., 1976), and Leonard J. Arrington and Davis Bitton, *The Mormon Experience: A History of the Latter-Day Saints* (New York: Alfred A. Knopf, 1979).

9. Hill, *loc cit.,* p. 110.

10. See, generally, Arrington and Davis, *The Mormon Experience,* and Gustive O. Larsen, "Government, Politics, and Conflict" and "The Crusade and the Manifesto," in Poll, *Utah's History.*

11. Lowry Nelson, *The Mormon Village: A Pattern and Technique of Land Settlement* (Salt Lake: University of Utah Press, 1952), p. xiii. "The Mormon's achievement in community building was a means to an end. The intense religious motivation of the group was the primary cause of success."

12. *Ibid.,* p. xv.

13. *Ibid.,* p. 28.

14. *Ibid.*

15. *Ibid.,* p. 277.

16. *Iron County Record,* Cedar City, Utah, October 16, 1946.

17. Albert Fisher, *Geography of Utah* (Salt Lake: University of Utah Press, 1980), pp. 20–25.

18. See Elroy Nelson, *Utah's Economic Patterns* (Salt Lake: University of Utah Press, 1956), pp. 3–10.

19. *Ibid.,* p. 6.

20. *Iron County Record,* Cedar City, Utah, January 1, 1948.

21. *Ibid.*

22. Nelson, *The Mormon Village,* p. 13.

23. *Ibid.,* p. 95.

24. *Ibid.*, p. 96.
25. *Ibid.*, pp. 96–97.
26. Fisher, *Geography of Utah*, pp. 20–21.
27. *Ibid.*, pp. 20–22.
28. *Ibid.*, pp. 22–23.
29. Gottlieb and Wiley, *America's Saints*, p. 184.
30. Michael Barone and Grant Ujifusa, *The Almanac of American Politics* (Washington, D.C.: National Journal, 1984), p. 1183.
31. "State Adjutant Given Plans for Home Guard," *Iron County Record*, Cedar City, Utah, August, 3, 1950.
32. *Iron County Record*, Cedar City, Utah, January 4, 1951.
33. *Washington County News*, St. George, Utah, March 1, 1951.
34. *Iron County Record*, Cedar City, Utah, May 21, 1953.
35. *Ibid.*
36. James B. Mayfield, *The First 80 Years: Utah Voting Behavior In Perspective*, 1980 (unpublished), p. 78.
37. *Ibid.*, p. 53.
38. *Ibid.*, p. 67.
39. *Ibid.*, Appendix B.
40. *Ibid.*
41. *Ibid.*, p. 77.
42. Gottlieb and Wiley, *America's Saints*, p. 70.
43. *Ibid.*, pp. 70–71.
44. *Ibid.*
45. See, generally, Nelson, *Utah's Economic Patterns*, sections V–VII.
46. See James L. Clayton, "Contemporary Economic Development," in Poll, *Utah's History*
47. Gottlieb and Wiley, *America's Saints*, p. 73.
48. F. Ross Peterson, "Utah Politics Since 1945," in Poll, *Utah's History*, p. 518.
49. *Ibid.*
50. *Ibid.* See stories and copy run in *Deseret News* on October 25, 26, 29, and 31, 1950; and November 6, 1950.
51. *Ibid.*, pp. 518ff. A glance at the results of the presidential elections in Utah indicates that the 1950 senatorial campaign was the watershed event in the state's move toward strong support for the Republican Party.

Two-Party Vote-Presidential Elections in Utah, 1932–1984

Presidential election	Democratic		Republican	
1932	Roosevelt:	57.9%	Hoover:	42.1%
1936	Roosevelt:	69.9	Landon:	30.1
1940	Roosevelt:	62.4	Wilkie:	37.6
1944	Roosevelt:	60.5	Dewey:	39.5
1948	Truman:	54.5	Dewey:	45.5
1952	Stevenson:	41.1	Eisenhower:	58.9
1956	Stevenson:	35.4	Eisenhower:	64.6
1960	Kennedy:	45.2	Nixon:	54.8
1964	Johnson:	54.7	Goldwater:	45.3
1968	Humphrey:	39.6	Nixon	68.4
1972	McGovern:	28.1	Nixon:	71.9
1976	Carter:	34.0	Ford:	62.0
1980	Carter:	21.0	Reagan:	73.0
1984	Mondale:	23.0	Reagan:	75.0

Compiled from Mayfield, *First 80 Years*, Appendix B, and Barone and Ujifusa, *Almanac of American Politics*, pp. 1185–1186.

52. *Ibid.*

53. *Ibid.*

54. *Ibid.*

55. *Health Effects of Low Level Radiation,* 1979 Hearings, Vol. II, p. 1417.

56. AEC, *Atomic Test Effects In The Nevada Test Site Region,* Washington, D.C., January, 1955, p. 1.

57. *Washington County News,* St. George, Utah, January 25, 1951.

58. *Deseret News,* January 27, 1951. A companion story, on page one, stated that "Vegas Blast Causes Little Stir."

59. "Atom Blast Startles Men on Road," *Deseret News,* January 28, 1951.

60. *Ibid.*

61. "Mt. Trumbull," *Washington County News,* St. George, Utah, February 1, 1951.

62. "Observations," *Washington County News,* February 1, 1951.

63. *Deseret News,* January 28, 1951. Readers were assured "that they are safe from the poisonous effects of an atom bomb which exploded with a thunderous roar in a remote valley northwest of Las Vegas before dawn Saturday."

64. *Ibid.*

65. "Vegans See Flash From 3rd A-Blast," *Deseret News,* February 1, 1951.

66. *Deseret News,* February 2, 1951.

67. *Ibid.*

68. "No Danger Seen For Utah In Radioactivity," *Deseret News,* February 2, 1951.

69. "A-Tests Spread Radioactive Snow Over East," *Deseret News,* February 3, 1951.

70. "Tuesday Atom Blast Finishes Test Runs," *Deseret News,* February 6, 1951.

71. "Atom Blasts Don't Worry Vegans Much," *Deseret News,* February 4, 1951.

72. *Deseret News,* January 29, 1951.

73. "Planning Service Offers New Home A-Bomb Shelter," *Deseret News,* February 3, 1951.

74. *Washington County News,* February 8, 1951.

75. *Deseret News,* January 28, 1951.

76. *Ibid.*

77. "Are The A-Bomb Tests A Hazard To Regional Structural Safety?", *Deseret News,* February 4, 1951.

78. *Deseret News,* February 3, 1951.

79. *Deseret News,* February 2, 1951.

80. *Deseret News,* January 30, 1951.

81. *Ibid.*

82. *Ibid.*

83. *Ibid.*

84. "Vegas Enjoys Respite From Flare of A-Bomb," *Deseret News,* February 4, 1951.

85. "Terrific A-Bomb Testing Blast Felt In St. George," *Washington County News,* November 1, 1951.

86. *Washington County News,* November 1, 1951.

87. *Iron County Record,* Cedar City, Utah, November 1, 1951, p. 29.

88. "Good and Bad News From Around The World," *Deseret News,* November 4, 1951. In an article, "Versatile A-Bomb Verified In Test," the *Deseret News* reported that America "has developed a tactical aerial atomic bomb that can blast enemy concentrations without harming friendly forces in foxholes little more than one-half a mile away."

89. *Washington County News,* April 3, 1952.

90. *Deseret News,* April 5, 1952.

91. *Deseret News,* April 17–21, 1952.

92. *Deseret News,* April 25, 1952.

93. *Deseret News,* May 9, 1952.

94. *Deseret News,* May 8, 1952.

95. *Ibid.*

96. *Deseret News,* May 9, 1952.

97. U.S. Department of Energy, *Operation Upshot-Knothole, 1953, U.S. Atmospheric Nuclear Warhead Tests, Personnel Review,* Washington, D.C.: DOD, DNA #6014F, 1982, pp. 2, 31.

98. *Iron County Record,* Cedar City, Utah, March 5, 1953.

99. "Utah Woman Sees A-Test," *Deseret News,* March 17, 1953.

100. "Eye Witness Account Given Of Yucca-Flats 'Atomic Device' Explosion," *Iron County Record,* March 19, 1953.

101. *Ibid.*

102. *Deseret News,* March 18, 1953.

103. *Deseret News,* March 26, 1953.

104. *Deseret News,* March 27, 1953.

105. *Ibid.*

106. *Health Effects of Low Level Radiation,* 1979 Hearings, Vol. II, p. 145.

107. *Ibid.,* p. 146.

108. *Deseret News,* March 27, 1953.

109. *Deseret News,* April 29, 1953.

110. *Deseret News,* May 1, 1953.

111. *Ibid.*

112. *Deseret News,* May 20, 1953.

113. *Ibid.*

114. *Deseret News,* May 21, 1953.

115. *Ibid.*

116. "A-Blasts and Weather," *Deseret News,* May 21, 1953.

117. *Iron County Record,* May 21, 1953.

118. *Iron County Record,* May 28, 1953.

119. *Iron County Record,* June 4, 1953.

120. *Iron County Record,* June 25, 1953.

121. *Ibid.*

122. *Ibid.*

123. *Bulloch v. U.S.,* 145 *F. Supp* 824 (1956); *Bulloch v. U.S.,* 95 *F Rules* 123 (1982).

124. "Effects Of Atom Blasts On Southern Utah Discussed By U of U Student," *Iron County Record,* May 7, 1953.

125. *Ibid.*

126. U.S. Department of Energy Dose Assessment Advisory Group Meeting, transcript, October 29, 1983, p. 219.

127. *Ibid.*

128. *Health Effects of Low Level Radiation,* 1979 Hearings, Vol. II, p. 157 (Commissioner Murray).

129. *Ibid.* (Dr. Dunning, AEC staff).

130. *Ibid.* (Commissioner Zuckert).

131. *Ibid.,* p. 161.

132. *Ibid.,* p. 167.

133. *Ibid.,* pp. 167–169.

134. *Ibid.,* p. 172.

135. *Ibid.*

136. *Ibid.*

137. *Washington County News,* February 17, 1955.

138. *Ibid.*

139. AEC Report, *Atomic Test Effects,* 1955, p. 1. At the same time, Dr. Allen C. Graves, AEC science Advisor at the Nevada Test Site, spoke in Salt Lake City on January 20, 1955 and assured Utahns that the AEC was taking "every possible precaution to prevent

inconvenience or danger to the public, including new procedures to minimize blast effects, and to reduce the fallout of radioactive material from the atomic clouds resulting from the explosions." Graves stated that "Preventing excessive fallout is first of all a matter of preventing dirt from the ground becoming mixed in the atomic cloud. This time we will use as many air bursts as possible, and, for the other explosions, will place the nuclear devices on 500-foot towers, higher than have ever been used before." In "Official Calms Utah Fears on '55 A-Blasts," *The Salt Lake Tribune,* January 21, 1955.

140. *Iron County Record,* February 24, 1955.

141. *Health Effects of Low Level Radiation,* 1979 Hearings, Vol, II, pp. 176–177.

142. *Ibid.,* p. 177 (Commissioner Nichols).

143. *Ibid.,* p. 178.

144. *Ibid.*

145. *Ibid.,* p. 179.

146. *Ibid.,* p. 180.

147. *Ibid.*

148. *Ibid.,* p. 181.

149. "A-Bomb Inquiry Explained By City Office," *Washington County News,* March 3, 1955.

150. *Deseret News,* March 8, 1955.

151. *Iron County Record,* March 10, 1955.

152. "Observations," *Washington County News,* March 24, 1955.

153. *Ibid.*

154. "Chamber of Commerce Hears Discussions, See A-Bomb Fallout Picture," *Washington County News,* April 14, 1955.

155. *Iron County Record,* March 3, 1955.

156. *Deseret News,* May 4, 1955.

157. *Deseret News,* May 13, 1955. An earlier *Deseret News* editorial, "Dreams or Nightmares," May 7, 1955, pleaded for an end to nuclear testing and said that "the more details one reads, the more one can be glad that the testing season is about over."

158. *Deseret News,* May 27, 1957.

159. Metzger, *The Atomic Establishment,* p. 83.

160. See *Iron County Record,* August 8, 1957.

161. *Iron County Record,* June 6, 1957.

162. *Deseret News,* May 30, 1957.

163. *Ibid.*

164. *Bulloch v. U.S.,* 145 F Supp 824 (1956).

165. *Deseret News,* October 8, 1957.

166. Quoted in Metzger, *Atomic Establishment,* p. 88.

167. "AEC Conducts Initial Tests of 1957 Summer Atomic Series," *Iron County Record,* May 30, 1957.

168. *Deseret News,* April 26, 1957.

169. *Deseret News,* April 27, 1957.

170. *Deseret News,* June 26, 1957.

171. On May 27–29 and June 3–7, 1957, the Special Committee on Radiation of the Joint Committee on Atomic Energy, 85th Congress, 1st Session, conducted hearings on *The Nature of Radioactive Fallout and its Effects on Man.* The hearings were the first ever held by Congress on this issue. Fifty scientific witnesses testified during the hearings on the subject of the "long-term radiation hazards, both from the military and peacetime atomic energy program" (p. 4). No less than 25 AEC commissioners and scientific staff persons testified before the committee. In addition to AEC commissioners present, the directors of the Argonne National Laboratory, the Brookhaven National Laboratory, Oak Ridge Laboratory, Los Alamos Scientific Laboratory, and the director of the AEC's Division of Biology and Medicine attended the hearings. Other leading University and gov-

ernment (HEW, NIH, PHS, NSF) scientists testified before the Committee. After analyzing over 2,000 pages of testimony and examining the origin and distribution of fallout, the biological effects of radiation, tolerance levels, and the health effects of past and future tests, the findings were inconclusive. There was some agreement among the testifiers about the improbability of a "clean" bomb being produced, and that "fallout was hazardous to a degree" (pp. 777–778). But the important question about the effect of fallout on human health was not fully answered. Was there a threshold? Was the health impact of low levels of radiation a linear one? The biological relevance of low-level radiation was still unknown, although many scientists disagreed with the AEC's contention that a radiation threshold existed below which human contact with radiation was "safe." The major, unresolved question of 1957 was whether or not a safe radiation level or threshold did indeed exist, below which there were no increases in the incidence of leukemias or cancers. In order to answer the threshold question, the Subcommittee recommended, at the conclusion of the hearings, that the AEC accelerate its research program on fallout. As a result, a five-year, research program was formulated within the AEC.

172. "Letters To The Editor," *Deseret News,* June 25, 1957.

173. "Letters To The Editor: Another Protest Against Bomb Tests," *Deseret News,* July 3, 1957.

174. "Letters To The Editor," *Deseret News,* July 22, 1957.

175. "Letters To The Editor," *Deseret News,* September 17, 1957.

176. *Deseret News,* October 8, 1957.

177. *Ibid.*

178. *Ibid.,* See Eugene England, "Can Nations Love Their Enemies: An LDS Theology of Peace" and Howard Ball, "The Nation State: Immorality and Violence," *Sunstone,* 7 (November/December, 1982), for a view and critique of the Mormon view of international relations.

179. *Deseret News,* October 8, 1957.

180. *Ibid.*

181. "President Prods Soviet on Parley," *The New York Times,* October 26, 1958.

182. "U.S. Ready To Halt Testing For Year," *The New York Times,* October 31, 1958.

183. "Geneva Nuclear Talks Adjourn With Major Snags Unresolved," *The New York Times,* December 19; 31, 1958.

184. "Failure At Geneva," *The New York Times,* December 19; 31, 1958.

185. "Last Nevada Test Balked By Deadline," *The New York Times,* October 31, 1958.

186. *Deseret News,* October 27, 1958.

187. *Deseret News,* October 29, 1958.

188. Metzger, *Atomic Establishment,* p. 104.

189. *Deseret News,* September 17, 1958.

190. "Letters To The Editor," *Deseret News,* October 1, 1958.

191. *Ibid.*

CHAPTER FOUR

1. See J. Harvie Wilkinson III, *From Brown To Bakke* (New York: Oxford University Press, 1980), and Howard Ball, Dale Krane, and Thomas P. Lauth, *Compromised Compliance: Implementation of the 1965 Voting Rights Act* (Westport, Connecticut: Greenwood, 1982).

2. See, for example, Richard Funston's *Constitutional Counter-Revolution? The Warren Court and The Burger Court: Judicial Policy-Making in Modern America* (New York: John Wiley and Sons, 1977).

3. See n. 1.

4. Carl J. Johnson, "Cancer Incidence in an Area of Radioactive Fallout, Downwind from the Nevada Test Site," *Journal of the American Medical Association,* 251 (January 13,

1984), p. 231. See also Joseph L. Lyon, et al., "Childhood Leukemias Associated with Fallout from Nuclear Testing," *New England Journal of Medicine,* 300 (1979), pp. 397–402. According to Mahlon Gates, Manager of the Nevada Test Site, 87 of the total 121 atomic shots (72%) fired by the AEC between 1951 and 1958 dropped radioactive fallout offsite. Testimony before the House Subcommittee on Oversight and Investigations of the Committee on Interstate and Foreign Commerce, April 23, 1979, pp. 9–10.

5. Charles R. Smart, Joseph L. Lyon and Harmon J. Eyre, eds., *Cancer in Utah, 1966–1977,* Report No. 3, Utah Cancer Registry, Salt Lake City, Utah, September, 1979, p. 1.

6. Charles R. Smart and Joseph L. Lyon, eds., *Cancer In Utah, 1957–1974,* Report No. 2, November, 1975, pp. 4, 80. Their Report No. 1, September, 1972, noted that there were 91.9 Utah deaths (per 100,000 population) in 1968, whereas the national U.S. average was 159.5 deaths per 100,000 population, p. 5.

7. In, *Health Effects of Low Level Radiation,* 1979 Hearings, Vol. II, p. 2195.

8. *Ibid.,* pp. 2779–2793.

9. *Ibid.,* p. 2784.

10. Quoted from transcript of the Special Town Meeting, conducted by U.S. Senator Orrin Hatch (R-Utah) in St. George, Utah, April 17, 1979, p. 98.

11. *Ibid.,* pp. 73–74.

12. *Ibid.,* p. 6.

13. *Ibid.,* pp. 32–33.

14. *Ibid.,* pp. 54–55.

15. *Ibid.,* p. 55.

16. *Ibid.,* p. 47.

17. *Ibid.,* p. 84.

18. *Ibid.,* p. 92.

19. Quoted in *National Catholic Reporter,* April 23, 1982.

20. Testimony before the Committee on Labor and Human Resources, *Radiation Exposure Compensation Act of 1981,* Hearings, pt. 2, 97th Congress, 2d session, April 8, 1982, pp., 20–25, esp. p. 24.

21. Bonnie Remsberg, "Beneath the Cloud," *Family Circle,* November 15, 1983, p. 76. See also Mrs. McEwen's testimony at the Special Town Meeting, St. George, Utah, April 17, 1979, pp. 70–73.

22. Special Town Meeting, St. George, Utah, April 17, 1979, p. 34.

23. *Ibid.,* p. 57.

24. *Ibid.,* p. 99.

25. Andrew Baum, Raymond Fleming, and Jerome Singer, "Coping with Victimization by Technological Disaster," *Journal of Social Issues,* 39 (1983), p. 123. See also his "Coping with Chronic Stress at Three-Mile Island," *Health Psychology,* 2 (1983), p. 151.

26. Baum, "Coping with Victimization," p. 125.

27. Laura Davis, Andrew Baum, and Daniel Collins, "Stress and Control-Related Problems at Three-Mile Island," *Journal of Applied Social Psychology,* 12 (1982), p. 351.

28. Roxanne L. Silver and Camille B. Wortman, "Coping with Undesirable Life Events," in Judy Gerber and Martin E. P. Seligman, eds., *Human Helplessness: Theory and Applications* (New York: Academic Press, 1980), p. 317.

29. Baum, "Coping with Victimization," p. 124.

30. Silver and Wortman, "Coping with Life Events," p. 317.

31. *Ibid.*

32. Raymond Fleming, Andrew Baum, Martha Gisriel, Robert Gatchel, "Mediating Influences of Social Support on Stress at Three-Mile Island," *Journal of Human Stress,* September (1982), p. 21. "Three-Mile Island residents reporting low levels of social support reported more symptoms, depression, anxiety and alienation, than did any other subjects," pp. 17–18.

33. *Ibid.,* p. 19.

34. Victor Frankl, *Man's Search for Meaning* (New York: Washington Square Press, 1963).

35. Silver and Wortman, "Coping with Life Events," pp. 332, 334.

36. Norman Solomon and Harvey Wasserman, "We All Live Downwind," *Environmental Action*, April (1983), p. 16.

37. Silver and Wortman, "Coping with Life Events," p. 337; Davidson, Baum, and Collins, "Stress at Three-Mile Island," p. 352. Baum discovered that at the time of the Three-Mile Island crisis, plant site officials gave out false information to the public.

38. Mrs. Irma Thomas, quoted on KUTV, "Downwind" Salt Lake City, December 17, 1983.

39. See Theodore White, *Breach of Faith: The Fall of Richard Nixon* (New York: Atheneum, 1975).

40. *Radiation Exposure Compensation Act of 1981*, Hearings, pt. 2, 1982, p. 21.

41. Silver and Wortman, "Coping with Life Events," p. 337.

42. Johnson, "Cancer Incidence Downwind," pp. 233–234.

43. *Ibid.*, p. 318.

44. Testimony in *Radiation Exposure Compensation Act of 1981*, Hearings, pt. 1, October 27, 1981, p. 131.

45. *Ibid.*, pp. 132, 135.

46. Interview, Janet Gordon with Katharine Greene, Research Associate, University of Utah, April, 1984.

47. Interview with author, April, 1984.

48. Testimony of Gloria Gregerson, member, board of directors, Citizen's Call, in *Radiation Exposure Compensation Act of 1981*, Hearings, pt. 2, 1982, p. 25.

49. *Ibid.*

50. *Ibid.*, p. 29.

51. *Ibid.*, p. 26.

52. See *Citizen's Voice*, Vol. 2, No. 2 (Fall 1983), p. 2.

53. Since 1952, a major radiation issue in Colorado has been the Rocky Flats plant contamination of the soil and air in and around the facility. Rocky Flats, Colorado is the location of a nuclear weapons plant run by Rockwell International for the U.S. Department of Energy. The plant manufacturers plutonium components which are used to trigger nuclear weapons detonations. With the discovery of an excessive number of brain tumors among Rocky Flats employees (eight observed deaths, with two expected), four studies were conducted by medical experts (between 1981 and 1984) at the Los Alamos National Laboratory in New Mexico (which is under the supervision of the U.S. Department of Energy.) All the studies confirmed the excess number of deaths but none linked the deaths to radiation exposure on the job. Governor Richard D. Lamm, Colorado, appointed a five-person, Rocky Flats Employees Health Assessment Group to review the results of the Los Alamos studies. The Governor's group concluded that there was no "cover up" or "scientific prostitution," but called for additional studies for the rest of the workers' lives by agencies and scientists not associated with the government or the company. At this time there is "no obvious explanation for a higher incidence of brain tumors" among the plant workers. Dr. Carl Johnson, a critic of the government, claimed that his studies of the workers "showed an eightfold excess of brain tumors, compared with the general male population; a threefold excess of malignant melanomas and an excess of respiratory cancer."

The basic question for the Governor's group was "whether there exists a toxic exposure which has significant, harmful, and detectable impacts upon human health." Simply put, were the brain tumors caused by "measurable external" low levels of radiation exposure. While the review group was puzzled about the fact that the incidence of brain tumors was higher than expected, "definitive answers now" are unavailable because there has not been a sufficient "passage of time and performance of autopsies." See Governor's

Science and Technology Advisory Council, *Report of the Rocky Flats Employees Health Assessment Group,* March 21, 1984, Denver, Colorado, pp. 1, 5, 9; "Study Says Flats Tumor Incidence Not Explainable," *Denver Post,* April 11, 1984; "Cover-Up Rejected in Flats Cancer Finding," *Rocky Mountain News,* April 11, 1984.

In a national precedent-setting ruling involving the Rocky Flats plant, the Colorado Industrial Commission, on April 24, 1984, ordered the Rocky Flats Nuclear Plant, run by Kerr-McGee, to pay $40,000 in compensation to Mrs. Florence Krumback. Her husband, LeRoy, had been an employee in the nuclear plant for over fifteen years and, one month after leaving his job, had died of cancer. "It's the first case in the country where a worker exposed to permissible levels of radiation proved his cancer was caused by that radiation. . . . It says something about the exposure guidelines," said the widow's attorney, Bruce H. DeBoskey. Krumback, 65, had died in January, 1974. The U.S. Department of Energy, which owns Rocky Flats, had opposed this kind of action by the Industrial Commission, arguing that federal radiation exposure limits were not exceeded. See "N-Plant Ordered to Pay Cancer Victim's Widow," *Salt Lake Tribune,* April 25, 1984.

The April 19, 1984 order, W.C. No. 2-923-974, *In Re LeRoy A. Krumback v. Dow Chemical Company and Travelers Insurance Company,* of the Colorado Industrial Commission, acknowledged the existence of "reasonable probability" that Krumback's death from colon cancer was due to radiation exposure he received at the Rocky Flat plant. The Commission concluded that the deceased's chronic exposure to ionizing radiation was the proximate cause of the cancer which resulted in his death. This board acted pursuant to a Colorado Court of Appeals order, in case No. 82 CA 1443, *Krumback v. Dow Chemical Company, et al.,* September 1, 1983. The state appellate court had reviewed an earlier order of the Industrial Commission (November 24, 1982) in the Krumback appeal, in which the board had ruled against the wife of the deceased, and set it aside.

The Court of Appeals ruled that the hearing officer's conclusions, set aside by the Commission in the November, 1982 ruling, had to be admitted in the compensation hearing. In the earlier hearings, the referee had accepted expert testimony from health physicists and had ruled that a causal connection existed between Krumback's exposure to radiation at the Rocky Flats plant and his subsequent death from colon cancer.

The Court set "reasonable probability" as the guiding standard in the case (pp. 2–3), allowed the expert testimony of the two health physicists to stand (pp. 4–6), and ordered the Industrial Commission to enter an order using the standard and the conclusions of the expert witnesses. The subsequent Industrial Commission ruling (April 19, 1984) following the appellate court remand, affirmed the referee's decisions and awarded the widow $40,000.

Note also the generic issue of plutonium exposure in the Karen Silkwood lawsuit against the Kerr-McGee Corporation in the federal courts. See generally, Howard Kohn, *Who Killed Karen Silkwood?* (New York: Summit Books, 1981). On January 11, 1984, the U.S. Supreme Court decided the case of *Silkwood v. Kerr-McGee Corporation,* 104 SCT 615 (1984). A state jury had awarded the estate of Karen Silkwood compensatory and punitive damages "arising out of the escape of plutonium from a federally-licensed nuclear facility (Kerr-McGee Nuclear Plant, Cimmaron, near Cresent, Oklahoma)" (p. 617). Justice White pointed out that Karen Silkwood was contaminated by plutonium during a three-day period (the autopsy revealed 8.8 nanocuries of plutonium in Karen's body), "which might have manifested itself in the form of lung cancer and chromosome damage at some future time" (p. 635, n. 2). She was killed in an automobile accident shortly thereafter (p. 618). The Supreme Court Justice also noted that Kerr-McGee "did not always comply with NRC regulations" (p. 618). The jury awarded her estate actual damages of $505,000 and punitive damages of $10 million. The U.S. District Court judge upheld the jury verdict but the U.S. Court of Appeals for the Tenth Circuit reversed the jury's award. Silkwood appealed and the U.S. Supreme Court reversed and remanded the case, holding that the award of punitive damages was not preempted by federal law. "Insofar as damages for radiation injuries are concerned, preemption (should not be interpreted as Congress occupying the field of safety so completely) that state remedies are foreclosed" (p. 626). See also "Karen Silkwood, Federalist," *Regulation,* 8 (January/February, 1984), pp. 4–7.

54. Jay Truman, "The View from Downwind," in *Testing News,* 1982.

55. Solomon and Wasserman, "We All Live Downwind," p. 17.

56. Truman, "View from Downwind."

57. Jay Truman, *A Guide to A.C.T.B.,* Salt Lake City, *Downwinders,* 1982, p. 6.

58. Solomon and Wasserman, "We All Live Downwind," p. 19.

59. *Ibid.,* p. 17.

60. Quoted on KUTV, "Downwind," Salt Lake City, December 17, 1983.

61. *Ibid.*

62. Bennie Levy, telephone interview with Research Associate Kathanne Greene, April, 1984.

63. *Ibid.*

64. *Radiation Exposure Compensation Act of 1981,* Hearings, pt. 1, 1981, pp. 155, 158.

65. Janet Raloff, "Compensating Radiation Victims," *Science News,* 124 (1983), p. 330.

66. *Radiation Exposure Compensation Act of 1981,* Hearings, pt. 1, 1981, pp. 132, 141.

CHAPTER FIVE

1. Quoted in James D. Watson, *The Double Helix* (New York: Atheneum, 1969), p. vii. The following description of DNA and the effects on the cell of chromosomal damage is from the *GAO Report to Congress: Problems in Assessing the Cancer Risks of Low-Level Ionizing Radiation Exposure,* Vol. I, January, 1981, pp. 22, 24:

> The DNA is the human cell's information molecule. DNA has four basic units called nucleotides containing cellular information. The DNA is double-stranded; . . . one strand of the DNA is a model for building the other strand. This arrangement is essential for cellular reproduction, and for repair of DNA damage. The two strands are twisted into a double helix. The DNA is further coiled and clumped so that 5 feet of it (one ten millionth of an inch wide) is contained in each human cell . . . DNA controls the biological machinery of a cell through complex, only partly understood mechanisms. Damage done to the DNA—whether by radiation, chemicals, or other physical processes can have profound effects on the cell. The cell might be unable to reproduce itself; or it might survive and transmit faulty instructions to its daughter cells, thereby producing cells with mutations. . . . Ionizing radiation mainly affects cells by damaging their DNA.

2. John Gofman, *Radiation and Human Health* (San Francisco: Sierra Club Books, 1981), p. 57.

3. Watson, *Double Helix,* p. 186.

4. Gofman, *Radiation,* pp. 57–58.

5. *Ibid.,* p. 58.

6. Saffer and Kelley, *Countdown Zero,* p. 268. Charles Moore's essay, "Radiation and Preconception Injuries," *Southwestern Law Review,* 28 (1974), maintains that this injury, occurring in genetic structure of a parent's DNA "before there is any separate existence whatsoever of the person," is probably "the most difficult and disturbing of the possible injuries" brought about by exposure to radiation. P. 415, n. 7; 416.

7. Gofman, *Radiation,* p. 54.

8. Wayne D. Le Baron, *The Reluctant Survivors* (Salt Lake City; Dream Garden Press, 1984), pp. 45–46.

9. Gofman, *Radiation,* p. 85, See also William Hendee, "Real and Perceived Risks of Medical Radiation Exposure," *Western Journal of Medicine,* 138 (March, 1983), pp. 380–386; GAO Report, *Problems Assessing Cancer Risks,* pp. 27–28.

10. *Ibid.*, p. 58; pp. 25–26, 32. While there is

some evidence that cells have mechanisms that can repair more extensive damage to the DNA, . . . the efficiency of the repair varies. . . . Repeated, low-level doses injure the DNA stem cells. An accumulation of small injuries to the stem cells increases the chance of a malignant change occurring in one stem cell. When a cell arises which is unresponsive to the usual regulatory signals, it continues to divide and floods the body with its progeny, it is called cancer/leukemia."

11. *Ibid.*, p. 105.

12. Gofman, *Radiation*, p. 412.

13. Raloff, "Compensating Victims," 1983, p. 412.

14. See, for example, the research recommendations made by a special task force for the Governor of Colorado, in *Report of the Rocky Flats Employees Health Assessment Group*, March 21, 1984, Denver, Colorado.

15. Jacob I. Fabrikant, "Epidemiological Studies on Radiation Carcinogenesis in Human Populations Following Acute Exposure: Nuclear Explosions and Medical Radiation," *Yale Journal of Biology and Medicine*, 54 (1981), p. 459.

16. *Ibid.*, p. 457.

17. See, for example, Charles E. Land's critique of Dr. Lyon's Utah study, "The Hazards of Fallout or of Epidemiological Research?", *New England Journal of Medicine*, 300 (February 22, 1979), p. 431.

18. Eighth Dose Assessment Advisory Group meeting, Vol. I, October 20, 1983, U.S. Department of Energy, p. 219ff.

19. Memo, William D. Schaffer, Chairman, U.S. Interagency Task Force on Compensation for Radiation-Related Illnesses, from the Schaffer Report, Washington, D.C., 1980, p. 13.

20. Dr. Ken Smith, Research Associate, Family and Consumer Medicine, interview with author, April 27, 1984.

21. Mervyn Susser, *Causal Thinking in the Health Sciences* (New York: Oxford University Press, 1973), pp. 4, 5, 8, 9.

22. *Ibid.*, p. 9.

23. Dr. Ken Smith, interview with author, April 27, 1984. Epidemiologists in this area of medical research "compare the incidence of cancer in a group with an unusual radiation exposure to a control group, i.e. people similar to the exposed group, but who have not been exposed to unusual levels of radiation," GAO Report, *Problems Assessing Cancer Risks*, Vol. 2, pp. 2–29.

24. *The Forgotten Guinea Pigs*, 1980 House Report., p. 15.

25. Letter, August 27, 1965; *Health Effects of Low Level Radiation*, 1979, Vol. II, p. 2250.

26. In *Health Effects of Low Level Radiation*, 1979, Vol. II, September 12, 1962 Memo, p. 1984.

27. *Ibid.*, September 13, 1962 Memo, pp. 2018, 2023.

28. Schaffer Report, *Compensation*, p. 27.

29. Marvin Rallison, Blown M. Dobbyns, F. Raymond Keating, Joseph E. Rall, Frank H. Tyler, "Thyroid Disease in Children," *American Journal of Medicine*, 56 (April, 1974), p. 457.

30. *Ibid.*, pp. 458–459.

31. *Ibid.*, p. 462.

32. Lyon's testimony, *Radiation Exposure Compensation Act of 1981*, p. 177.

33. Marvin Rallison, press release, University of Utah, *Fallout Study*, July, 1982, p. 3.

34. *Ibid.*, p. 4.

35. *Health Effects of Low Level Radiation*, 1979, Vol. I, p. 363.

36. *Ibid.*

37. Schaffer Report, *Compensation*, p. 27.

38. *Radiation Exposure Act of 1981,* pp. 174–175, 178.

39. *Ibid.,* pp. 174, 178.

40. *Ibid.,* p. 175.

41. Joseph L. Lyon, Melville Klauber, John W. Gardner, and King Udall, "Childhood Leukemias Associated with Fallout from Nuclear Testing," *New England Journal of Medicine,* 300 (February 22, 1979), pp. 397–399.

42. *Ibid.,* p. 398.

43. *Ibid.,* p. 401.

44. Gofman, *Radiation,* p. 55.

45. *Health Effects of Low Level Radiation,* 1979, Vol. I, p. 375.

46. *Ibid.*

47. John Cairns, Joseph L. Lyon, Mark Skolnick, "Cancer Incidence in Defined Populations," in *Banbury Report No. 4* (New York: Cold Spring Harbor Laboratory, 1980).

48. *Ibid.,* p. 146.

49. *Ibid.,* p. 157.

50. Land, "Hazards of Fallout" p. 431; Charles E. Land, et al., "Childhood Leukemias and Fallout from the Nevada Nuclear Tests," *Science,* 223 (January 13, 1984), pp. 139–144.

51. Land, "Hazards of Fallout, p. 432.

52. *Ibid.,* p. 433.

53. Land, "Childhood Leukemias and the Nevada Nuclear Tests, p. 139.

54. *Ibid.,* p. 144.

55. *Ibid.*

56. *Ibid.*

57. *Ibid.,* p. 142.

58. *Ibid.,* pp. 142–143.

59. *Ibid.,* p. 144.

60. Saffer and Kelly, *Countdown Zero,* p. 152.

61. *Ibid.,* p. 153.

62. *Ibid.,* pp. 153–154.

63. Glyn G. Caldwell, Delle B. Kelley, and Clark W. Heath, Jr., "Leukemia Among Participants in Military Maneuvers at a Nuclear Bomb Test," *Journal of the American Medical Association,* 244 (October 3, 1980), pp. 1575–1578.

64. *Ibid.,* p. 1577.

65. *Ibid.,* p. 1578.

66. Schaffer Report, *Compensation,* p. 30.

67. V. P. Bond and L. D. Hamilton, "Leukemia in the Nevada Smoky Bomb Test," *Journal of the American Medical Association,* 244 (October 3, 1980), p. 1610.

68. *Ibid.*

69. *Ibid.*

70. Joseph L. Lyon and John W. Gardner, "Radiation Exposure and Cancer," *Journal of the American Medical Association,* 246 (November 13, 1981), pp. 2153–2154.

71. *Ibid.,* p. 2153.

72. *Ibid.*

73. *Ibid.* See also, Dr. Joseph L. Lyon, Associate Professor, Family and Consumer Medicine-Epidemiology, interview with author, April 30, 1984.

74. *Ibid.,* p. 2154.

75. Glyn G. Caldwell, Delle Kelley, Matthew Zack, Henry Falk, and Clark Heath, Jr., "Mortality and Cancer Frequency Among Nuclear Test (Smoky) Participants, 1957–1979," *Journal of the American Medical Association,* 250 (August 5, 1983), pp. 620–624.

76. *Ibid.,* p. 623.

77. *Ibid.,* pp. 621–622.

78. *Ibid.,* p. 622.

79. *Ibid.*, p. 623.

80. *Ibid.*, p. 624. It is noteworthy that Dr. Caldwell now sits on the Dose Assessment Advisory Group, Department of Energy, and another author, Dr. Heath, who still works for the Center for Disease Control, was a medical expert witness for the defendant, the U.S. Government, in the *Allen* case.

81. Harold L. Beck and Philip W. Krey, "Radiation Exposures in Utah from the Nevada Nuclear Tests," *Science,* 220 (April 1, 1983), pp. 18–24.

82. *Ibid.*, p. 18.

83. *Ibid.*, p. 22.

84. *Ibid.*

85. *Ibid.*, p. 23. Based on governmental estimates of doses received by the downwind people—about 60,000 person rads—the 1979 *Biological Effects of Ionizing Radiation* (BEIR) *Report* estimates that among the 172,000 downwind residents there may eventually occur between 18 to 48 cancers, of which 6 to 16 should be fatal. Given the plus or minus factor of 2 variability of these estimates, the BEIR range of excess cancers for the downwind communities would be 36 to 96 occurrences, of which 12 to 32 cancers would be fatal. Schaffer Report, *Compensation,* pp. 31ff.

86. *Ibid.*, p. 24.

87. *Ibid.*

88. Johnson, "Cancer Incidence Downwind," pp. 230–236.

89. *Ibid.*, p. 230.

90. *Ibid.*, p. 232.

91. *Ibid.*, pp. 233–234.

92. *Ibid.*, p. 235.

93. *Ibid.*, p. 236.

94. Lyon et al., "Childhood Leukemias Associated with Fallout," pp. 397–402.

95. Rallison, press release, University of Utah, *Fallout Study,* July, 1982, p. 2.

96. *Ibid.*

97. McDonald E. Wrenn, "A Review of External Radiation Dose Estimates Offsite Associated with Nuclear Testing at the NTS and Current Research," Paper presented at the 16th Annual Meeting of the NCRP, April, 1980, p. 180.

98. Hiroo Kato, et al., "Studies of the Mortality of A-Bomb Survivors," *Radiation Research,* 90–91; (1982), pp. 243–264; 395–432.

99. *Ibid.*, p. 396.

100. *Ibid.*, p. 249.

101. Dr. Joseph L. Lyon, interview with author, April 30, 1984.

102. *Ibid.*

103. *Ibid.*

104. *Ibid.*

105. F. Peter Libassi, former General Counsel, U.S. Department of Health, Education, and Welfare, stated that we might be underestimating the cancer-causing potential of low-level radiation by a factor of ten! "Recent studies of populations exposed to low levels of radiation suggest that the risks may be higher than earlier predictions." *Health Effects of Low-Level Radiation,* Vol. I, pp. 205, 270.

106. Yalow, "Reappraisal," p. 49.

107. Dr. Joseph L. Lyon, interview with author, April 30, 1984.

CHAPTER SIX

1. *Bulloch v. U.S.,* 145 *F. Supp* 824 (1956).

2. Bill Curry, "U.S. Ignored Atomic Test Leukemia Link," *Washington Post,* January 8, 1979. See also, Gordon Eliot White's articles in the *Deseret News:* "Leukemia Reports

Never Published," November 13, 1978; "War Fears Forced Nevada A-Site," December 8, 1978; "65 Nodule Report Released Only After a Fight," January 30, 1979; "Fallout Washed from Utah Cars," February 8, 1979.

3. Titus, "Government Responsibility for Victims of Atomic Testing," p. 282.

4. See House Subcommittee on Health and the Environment of the House Committee on Interstate and Foreign Commerce, *Effects of Radiation on Human Health—Health Effects of Ionizing Radiation,* Hearings, 95th Congress, 2d session, 1978;

House Subcommittee on Oversight and Investigations, Committee on Interstate and Foreign Commerce, *Health Effects of Low Level Radiation,* Hearings, 96th Congress, 1st session, April, May, August, 1979;

Senate Committee on Veterans Affairs, *Veterans' Claims for Disabilities from Nuclear Weapons Testing.* Hearings, 96th Congress, 1st Session, June 20, 1979;

Senate Subcommittee on Health and Scientific Research, Committee on Labor and Human Resources, and Committee on the Judiciary, *Radiation Exposure Compensation Act of 1979,* Joint Hearing, 96th Congress, 2d session, June 10, 1980.

House Subcommittee on Oversight and Investigations, Committee on Interstate and Foreign Commerce, *The Forgotten Guinea Pigs: A Report on Health Effects of Low Level Radiation Sustained as a Result of the Nuclear Weapons Testing Program Conducted by the United States Government.* 96th Congress, 2d session, August, 1980;

Senate Committee on Labor and Human Resources, *Radiation Exposure Compensation Act of 1981,* Hearing, 97th Congress, 1st session, October 27, 1981.

5. *The Forgotten Guinea Pigs,* 1980 House Rept., pp. 21–22.

6. Titus, "Government Responsibility," p. 291.

7. *Health Effects of Low-Level Radiation,* 1979 Hearing, Vol. 1, p. 25.

8. *Health Effects of Low-Level Radiation,* Vol. 2, p. 2443ff.

9. Joseph A. Califano, Jr., *Governing America: An Insider's Report from the White House and the Cabinet* (New York: Simon and Schuster, 1981), p. 421.

10. *Health Effects of Low-Level Radiation,* 1979 Hearing, Vol. 1, pp. 25–26.

11. Governor Scott Matheson, interview with author, May 10, 1984.

12. Memo, Craig Carlisle to Stewart Udall, May 15, 1981, p. 23.

13. Palma Strand, "The Inapplicability of Traditional Tort Law Analysis to Environmental Risks: The Example of Toxic Waste Pollution Victim Compensation," *Stanford Law Review,* 35 (February, 1983), p. 576.

14. *Ibid.,* p. 575.

15. *Ibid.*

16. Richard Delgado, "Beyond Sindell:' Relaxation of Cause-in-Fact Rules for Indeterminate Plaintiffs," *California Law Review,* 70 (1982), p. 583.

17. *Ibid.,* p. 883, n. 13.

18. *Ibid.,* p. 883, n. 12.

19. *Ibid.,* p. 883.

20. Gordon E. White, "Attorney Calls Cancer Victims War Casualties," *Deseret News,* December 23, 1978.

21. "The principal reason for limiting the scope of governmental tort liability, however, is the principle of separation of powers inherent in our governmental system." Mary F. Wyant, "The Discretionary Function Exception to Government Tort Liability," *Marquette Law Review,* 61 (1977), p. 166.

22. Debra Sholl, "The Nevada Proving Grounds: An Asylum for Sovereign Immunity," *Southwestern University Law Review,* 12 (1981), p. 631. See also Christine M. Doyle, "Government Liability for Nuclear Testing Under FTCA," *University of California, Davis, Law Review,* 15 (1982), p. 1007.

23. *Feres v. U.S.,* 340 US 135 (1950).

24. Allan Favish, "Radiation Injury and the Atomic Veteran," *Hastings Law Journal,* 32 (1981), pp. 163–164.

25. *Dalehite v. U.S.,* 346 US 15 (1953).

26. Sholl, "Nevada Proving Grounds," p. 638. See also Murphy, "Atomic Energy and the Law," *Vanderbilt Law Review,* 12 (1958), p. 229, for an excellent, early bibliography on this subject.

27. Doyle, "Government Liability," pp. 1007–1008.

28. Sholl, "Nevada Proving Grounds," p. 635 (n. 42). Norman B. Antin "Constitutional Tort Remedies," *Connecticut Law Review,* 12 (1980), p. 494, suggests that "inequitable decisions still permeate the case law as a consequence of an immunity doctrine which remains unnecessarily broad."

29. Over the years, given *Dalehite*'s precarious precedential perch and the operation/ planning distinctions developed in it, it has been somewhat eroded by subsequent Supreme Court decisions and the lower federal court judgments. The federal judges have "struggled repeatedly to formulate a principled analysis of government conduct that will effectively guide the application of the "discretionary function" exception . . . while remaining consistent with the broad language of *Dalehite*." [*Allen v. United States,* 527 F Supp, 476 (1981), p. 479.] Indeed, some federal circuit courts have discarded the *Dalehite* language "in favor of an approach *looking to the policies underpinning* the discretionary function exception *and to the nature of the discretion involved in each case.*" In *Smith v. United States,* 375 F2d 243 (1967), the Fifth Circuit concluded that the *Dalehite* distinction between planning and operations was "specious" and, if strictly followed by the courts, would sap the "corpuscular vitality" of the Federal Torts Claims Act.

30. See also *Rayonier v. United States,* 352 US 315 (1957). This was an important modification by the justices: after a planning judgment was made, the government had to ensure that the operation of the activity was done properly or else the agents and the government would be subject to suits under the Federal Tort Claims Act. That same year, in a lower federal court opinion, *United States v. Union Trust Company,* 221 F2d 62 (1955) [affirmed by the U.S. Supreme Court, 350 US 907 (1955)], the federal court held that while "discretion was exercised to operate the [aircraft control] tower," the "tower personnel had no discretion to operate it negligently."

In a 1957 case involving national defense and nuclear testing, *Bartholomae v. United States,* 253 F2d 716, the Ninth U.S. Circuit Court of Appeals concluded (1) that decisions in these areas were "authorized at the highest levels of government for the public benefit"; (2) that all important judgments, from preparation to final detonation, were made at the highest levels of government; but (3) that there is a "duty to act with due care" in implementing the decision to conduct tests. Although there is the discretionary function immunity for governmental planning, "when the government ignores minimum requirements of ordinary care in the performance of a discretionary function, it may be liable for resulting damages." Sholl, "Nevada Proving Grounds," p. 641.

In a 1964 Ninth U.S. Circuit Court of Appeals decision, *United Airlines v. Weiner,* 335 F2d 379, the federal appeals court drew up a catalogue of discretionary and operational activities. Firefighting was a discretionary judgment; negligent conduct by the federal firefighters was not immune from liability. It was a discretionary judgment to admit a patient to a federal hospital; negligent treatment of patients was an operational activity subjecting the government to tort suits. It was a discretionary judgment to establish a U.S. Post Office at a particular location, but it was negligent operational action to fail to install handrails for handicapped people. Discretion is involved in establishing air traffic control towers at airports; negligent conduct of the controllers is not given immunity by the Federal Tort Claims Act. It was a discretionary function to reactivate a military base, but it was operational negligence not to construct a drainage system. In sum, the Court was limiting "discretionary function" Federal Tort Claims Act exceptions to the *initial planning decision,* for example, to build an air control tower. It was also stressing the point, made in earlier federal cases, that once the decision was made to control air traffic, the operations of the tower were subject to Federal Tort Claims Act suits, if negligent acts or omissions by air controllers led to injury, loss of property, or death.

In 1973, the U.S. Supreme Court, in the case of *Doe v. McMillan* 412 *US* 306, indicated that, in "discretionary function" cases, a pragmatic, judicial balancing process is at work between private and public interests: the harm the individual suffered against the threat to effective government. When a case focused on a congressional subcommittee's release of information (which was sent back for further examination), the Court majority indicated that balancing had to take place by the federal judge.

In *Ozark Airlines v. Delta*, 402 *F Supp* 687 (1975), a federal district court judge in Illinois continued the clarification of *Dalehite*. When the government undertakes to perform services not required by specific legislation, "it has the duty to perform these services carefully."

In a 1977 case involving a military doctor's negligence in the treatment of a female dependent, *Jackson v. Kelley*, 557 *F2d* 735 (1977), the U.S. Tenth Circuit Court of Appeals reinterpreted the *Dalehite* "discretionary function" doctrine. Chief Judge Lewis wrote about the need for a "careful inquiry" because of the "competing policies underlying official immunity from money damages";

> From the perspective of the individual citizen, some compensatory remedy must be available to vindicate injury inflicted by government officials. From the perspective of the public interest in the effective administration of policies of government, officials must be free to exercise their duties, especially discretionary duties, without having to defend their actions in court.

However, in this judicial balancing of individual demands for justice and the public's desire for innovative administrators, the "effective administration of policies of government is not impaired if officials with ministerial duties are answerable in damages for failure to perform obligatory functions with reasonable care":

> A duty is discretionary if it involves judgment, planning or policy decision. It is not governmental discretion if it involves enforcement or administration of a mandatory duty at the operational level, even if professional, expert evaluation is required. The key is whether the duty is mandatory or whether the act complained of involved policy-making or judgment . . . [The doctor's activity was] purely of a medical nature and is not protected.

31. *Allen*, at p. 107.

32. Philip J. Cooper, "Hard Judicial Choices," paper presented at the 1982 South Eastern Conference of Public Administrators meeting, November, 1982, pp. 13–14.

33. *Radiation Exposure Compensation Act of 1981*, Part II, p. 239.

34. *Ibid.*, p. 241.

35. Samuel D. Estep, "Radiation Injuries and Statistics," *Michigan Law Review*, 59 (1960), p. 274.

36. Stephen M. Soble, "A Proposal for the Administrative Compensation of Victims of Toxic Substance Pollution: A Model Act," *Harvard Journal on Legislation*, 14 (1977), p. 706.

37. *Ibid.*, p. 707.

38. Note, "Tort Actions for Cancer," *Yale Law Journal*, 90 (1981), pp. 840, 848; Laura Treadway, "When a Vet 'Wants' Uncle Sam," *American University Law Review*, 31 (1982), p. 1095.

39. Sholl, "Nevada Proving Grounds," p. 628 n. 5.

40. Adolpho Franco, "Wollman v. Gross: Statute of Limitations and the FTCA," *Creighton Law Review*, 15 (1982), pp. 1073–1074.

41. Moore, "Radiation and Preconception Injuries," p. 416. See also Samuel D. Estep, "Radiation Injuries and Statistics," *Michigan Law Review*, 59 (1960), p. 265.

42. *Ibid.*, p. 417.

43. Note, "The Application of the Statute of Limitations to Actions for Tortuous Radiation Exposure," *Alabama Law Review*, 31 (1980), p. 509.

44. *Radiation Exposure Compensation Act of 1981*, Part II, p. 244.

45. Frank M. Johnson, "The Role of the Federal Courts in Institutional Litigation," *Alabama Law Review,* 32 (1981), p. 274.

46. Dorothy See, "Committee Organized to Aid Fallout Victims," *Deseret News,* October 23, 1978.

47. See, for example, Jenkins' decision in 1982 cable decency law case, *Home Box Office (HBO) v. Wilkinson,* 531 F Supp 986, 1982.

48. See generally, Joe Bauman, "Nominee Announced by White House," *Deseret News,* August 29, 1978; Frank Hewlett, "Bruce Jenkins Nominated to Post as Federal Judge," *Salt Lake Tribune,* August 29, 1978.

49. See Sheldon Goldman, "A Profile of Carter's Judicial Nominees," *Judicature,* 62 (November, 1978).

50. See Appendix A for a discussion of *Bulloch.*

51. Frank Hewlett, "Judgeship? Jenkins In Review," *Salt Lake Tribune,* May 27, 1978.

52. Bauman, "Nominee Announced," *Deseret News,* October 23, 1978.

53. "Judge Airs 'Human' Side of Judiciary," *Salt Lake Tribune,* December 7, 1978.

54. Bruce S. Jenkins, "Remarks Before Joint Meeting," *Utah Bar Journal,* 10 (January, 1982), p. 6.

55. Bruce S. Jenkins, "Communication in the Courtroom: The Lawyer as an Educator," *American Bankruptcy Law Journal,* 48 (1974), p. 377.

56. *Ibid.,* p. 377.

57. *Ibid.,* p. 377–378.

58. *Ibid.,* p. 382.

59. *Ibid.,* p. 385.

60. *Ibid.,* p. 384.

61. Jenkins, "Remarks," p. 4.

62. Fuller, *Day,* pp. 256–257.

63. *Allen* trial transcript, Vol. 13, October, 1982.

64. *Ibid.*

65. Leon Green, "The Duty Problem In Negligence Cases," pt. I, *Columbia Law Review,* 28 (1978), pp. 1015, 1022.

66. *Ibid.,* pt. II, *Columbia Law Review,* 29 (1979), p. 256.

67. See *Kuhne v. United States,* 267 F Supp 649 (1967).

68. Frank M. Johnson, Jr., "Federal Judges: Policymakers," *Utah Forum,* 1: (1981), p. 15.

69. Johnson, "Role of the Federal Courts," p. 274. On Friday, May 4, 1984, two Nevada Test Site workers' widows, who had unsuccessfully sued the government after their husbands were exposed to radiation from the *Baneberry* venting, asked Judge Foley to allow them to reopen the case to present new evidence from a medical doctor who was unable to testify at the earlier trial. See "Baneberry Widows Want Wrongful Death Case Reopened," *Las Vegas Review Journal,* May 5, 1984.

70. Jenkins, "Remarks," p. 4.

71. Green, "Duty Problem in Negligence Cases," p. 1022.

CHAPTER SEVEN

1. Gordon Dean, "The Impact of the Atom on Law," *University of Pittsburgh Law Review,* 128 (1951), p. 517.

2. Udall's testimony, *Radiation Exposure Compensation Act of 1981,* p. 264.

3. *Ibid.,* p. 262.

4. *Ibid.,* p. 264.

5. All quotes in this segment are from the *Allen v. United States* documents filed by plaintiffs and defendants in federal district court between 1979 and 1982.

6. See, generally, Prosser, *Law of Torts,* 2d ed.

7. See, for example, the most recent and major "Agent Orange" class action settlement, worked out at 4 a.m., May 7, 1984, the morning of the first day of the scheduled trial, in U.S. District Court in New York, *In Re Agent Orange,* MDL, No. 381 (ED, New York, January 19, 1979). This was the latest, although not the final episode in a lengthy, controversial legal/political struggle that began with military use of 12 million gallons of herbicides 2, 4, 5 — T and 2, 4 — D (Agent Orange, containing dioxin) in the jungles of Vietnam, in 1965–70. Banned in 1970 by the United States because it was associated with birth defects, the herbicide was forgotten until 1978, when Paul Reutershan, a Vietnam vet suffering from liver cancer, brought a $10 million suit into federal court against the government and the makers of Agent Orange. He charged that his cancer was due to his Agent Orange exposure in Vietnam. Although he soon died, class actions against both the government and the chemical companies were instituted across the country, in New York, Philadelphia, Chicago, Louisiana, and Mississippi. In addition, the chemical companies, led by Dow Chemical, brought suit in 1980 against the government, charging that the United States misused Agent Orange in Vietnam and was responsible for these disabilities. Arguing that they and their offspring were injured by the herbicide, Australian Vietnam veterans also sued the U.S. Government and U.S. chemical companies in 1980. In 1982, a federal judge in New York ruled that the makers of Agent Orange cannot be held liable if they can prove that the United States, upon ordering the use of Agent Orange, was aware that the chemical was hazardous to human life. A year later, the judge ruled that Dow Chemical and other chemical companies had to stand trial in what was then a 4-year-old suit, filed by 20,000 veterans.

On May 7, 1984, the impaneling of jurors began. Early that morning, a settlement was reached to avoid the long-scheduled trial. The chemical companies agreed to set up a $180 million package, a superfund, which will grow to $250 million in 6 years. "The surprise settlement, announced in Brooklyn's Federal District Court, brought to an abrupt end more than five years of tortious litigation," wrote Ralph Blumenthal in the *New York Times,* "Veterans Accept $180 Million Pact on Agent Orange," May 8, 1984, p. 1A.

The government's liability, however, was not removed by this settlement between the veterans and the chemical companies. Both groups, the vets and the companies, can now file claims against the government. See William C. Rempel, "Both Sides Target U.S. on Agent Orange," *Salt Lake Tribune,* May 9, 1984.

On May 9, 1984, following the settlement, a *New York Times* editorial raised obvious legal and moral questions: "But if society truly means to dispense mercy, there's a strange absence among the defendants in the Agent Orange case. Devastating crops and forests in Vietnam was government's idea, and the government's moral duty to the veterans transcends that of the chemical companies that supplied the herbicide. Whether or not Agent Orange is a real cause of disease, it has become a metaphor for the veterans' sufferings."

(The Office of Management and Budget has warned the Reagan Administration that "compensation for such exposure [Agent Orange, radiation, asbestos, etc.] could take on huge proportions and become a costly item in the federal budget for years to come," Robert Pear, "U.S. Liability Not Resolved by Veterans Suit Settlement," *New York Times,* May 8, 1984.)

8. Udall's testimony *Radiation Exposure Compensation Act of 1981,* Hearings, p. 264.

9. See Udall's testimony. Senate Subcommittee on Health and Scientific Research, Committee on Labor and Human Resources, and Committee on the Judiciary, *Radiation Exposure Compensation Act of 1979.* Joint Hearing, 96th Congress, 2d session, June 10, 1980, p. 72.

10. *Radiation Exposure Compensation Act of 1981,* p. 263.

11. *Ibid.,* p. 264.

12. Memo, Chris Cannon to attorneys for plaintiffs associated with the radiation cases, June 28, 1979, p. 1; Memo, Chris Cannon to Stewart Udall, July 18, 1979, p. 1.

13. *Ibid.,* July 18, 1979 Memo, p. 2.

14. *Ibid.,* p. 5.

15. *Ibid.,* p. 6.

16. *Ibid.,* pp. 10–11.

17. Memo, Craig Carlile to Stewart Udall, May 15, 1981, p. 3.

18. Henry Gill, interview with author, February 1, 1984, Washington, D.C.

19. *Ibid.*

20. Defendant's response, *Allen v. United States,* October 30, 1979, p. 5.

21. Defendant's response to plaintiff's request for information, *Allen v. United States,* November, 1979, p. 4.

22. Janet Gordon, interview with author, May 14, 1984.

23. Judge Bruce S. Jenkins, interview with author, May 9, 1984.

24. Memorandum Opinion, *Irene Allen et al. v. United States,* Civil No. C-7a-055-J, May 10, 1984, U.S. District Court, Utah, Jenkins, Judge, p. 1.

25. at 8–9.

26. at 9.

27. at 147.

28. at 148.

29. at 150.

30. at 152.

31. at 154.

32. at 159.

33. at 178.

34. at 181ff.

35. at 184.

36. at 186.

37. at 195–196.

38. at 199.

39. at 200.

40. at 241.

41. at 242ff.

42. at 246.

43. at 261.

44. at 306–307.

45. at 315.

46. at 316–317.

47. at 318.

48. *Ibid.*

49. at 319.

50. at 321.

51. at 319.

52. at 339.

53. at 343–344.

54. at 344, 374.

55. at 346, 374.

56. at 355.

57. at 345.

58. at 346.

59. at 324.

60. at 374.

61. at 407.

62. at 415–416.

63. at 417.

64. at 418.

65. Robert E. Keaton, "Creative Continuity in the Law of Torts," *Harvard Law Review,* 75 (January, 1962), p. 472.

66. V. Wayne Thode, "Tort Analysis: Duty-Risk v. Proximate Cause and the Rational Allocation of Functions between Judge and Jury," *Utah Law Review,* 1977 (1977), p. 8.

67. Strand, "The Inapplicability of Traditional Tort Analysis to Environmental Risks," *Stanford Law Review,* 35 (1983), p. 612.

68. *Ibid.,* p. 618.

69. Delgado, "Beyond *Sindell:* Relaxation of Cause-In-Fact Rules for Indeterminate Plaintiffs," p. 888.

70. *Ibid.,* p. 897.

71. Quoted in *Deseret News,* May 12, 1984.

72. Quoted in *Salt Lake Tribune,* May 11, 1984.

73. Quoted in *Deseret News,* May 12, 1984.

74. Remarks of Richard K. Willard, Acting Assistant Attorney General, Civil Division, U.S. Department of Justice, Washington, D.C., press release, November 8, 1984, p. 7.

75. *Ibid.,* pp. 7–8.

76. *Ibid.,* p. 8.

77. *Ibid.,* p. 9.

78. *Ibid.*

79. *Ibid.*

80. *Ibid.,* p. 10–11.

81. Schaffer Report, *Compensation,* p. 47.

82. See Saffer and Kelly, *Countdown Zero,* for a full discussion of the events at Camp Desert Rock, Nevada; See also Uhl and Ensign, *G.I. Guinea Pigs;* Dr. Dorothy Legarreta, Coordinator, Genetic Defects In Atomic Veterans, testified before Hatch's Committee in April, 1982, that she was finding "about 20% of the vets who report problems also have reported genetic defects in their children—organ defects, mental retardation, etc." *Radiation Exposure Compensation Act of 1981,* Vol. 2, p. 176.

83. See Iver Peterson, "Utah Decision May Be Only the Opening Round," *New York Times,* May 13, 1984.

84. See, generally, Freeman, *Nuclear Witnesses,* pp. 137–170; testimony by Navajo representatives, *Radiation Exposure Compensation Act,* 1981 Hearing, pt. 2, pp. 129–146. The uranium miners litigation is a particularly tragic story of government negligence that has evidently placed thousands of uranium miners' lives in jeopardy due to lung cancer. Apparently, Navajo lands in the Four Corners area of the Southwest contain rich uranium deposits. The AEC chose not to regulate the underground mining, and it was not until 1972 that federal radiation-level standards were established. When Kerr-McGee and other private corporations took over the mining operation, they found no AEC safety or health regulations, and an abundant supply of cheap labor. Dramatic increases in lung cancer were noted after 1965 by medical researchers working at NIOSH. Water was contaminated after coming in contact with the uranium in the mines. Radiation mill tailing ponds and piles now litter the Four Corners landscape, endangering humans and livestock. Compounding the litigation problem is the fact that Navajo tribal customs do not allow autopsies on tribal members. "Therefore, the only evidence on the information we can get is from the doctors who examined them while they were alive" (p. 130). A Navajo, Perry H. Charley, summed up the tragedy when he testified in 1981: "I have seen and felt my people's anguish. I have seen them gradually decline in health and have seen them dying or die" (p. 142).

85. David D. Davis, *Energy Politics* (New York: St. Martin Press, 1974), p. 152.

86. *Ibid.*

87. Domenici's testimony, *Radiation Exposure Compensation Act of 1979,* 1980 Hearings, p. 19.

88. See *Sylvia Barnson, et al. v. Foote Mineral Col, Vanadium Corporation and United States,* Civil No. 80-119 (U.S. District Court, Utah, Anderson, Judge). (Consolidation of four suits brought into Utah's U.S. District Court.) Judge Anderson, in a July 1985 ruling that "shocked" the plaintiff's attorney, Stewart Udall, issued a Summary Judgement to the U.S. His 13-page order dismissing the FTCA suit concluded that the government was immune from negligence claims of the uranium miners' survivors. He noted that, under the FTCA, "discretionary acts are immune, regardless of whether they are negligent." Said Udall:"We will of course appeal." See Joan O'Brien, "U.S. Immune From Mine Suits," *Salt Lake Tribune,* July 2, 1985, B-2.

89. See *John N. Begay, et al. v. United States,* Civil No. 80-982 (U.S. District Court, Copple, Judge). Union leader Tony Mazzochi's statement is quoted in Freeman, *Nuclear Witnesses,* p. 168.

90. Quoted in Freeman, *Nuclear Witnesses,* p. 144.

91. Quoted in *Deseret News,* May 12, 1984.

92. Quoted in *Deseret News,* May 13, 1984.

93. *Deseret News,* May 12, 1984.

94. *Ibid.*

95. Dr. Ron Preston, interview with Research Associate Ken Verdoia, Washington, D.C., May, 1984.

96. *Ibid.*

97. *Deseret News,* May 12, 1984.

98. Strand, "Inapplicability of Traditional Tort Analysis," p. 618.

99. *Ibid.*

100. In the recent Agent Orange settlement, another U.S. Federal District court judge, Jack B. Weinstein, stretched the law to make it responsive to the unusual circumstances of the dioxin case. The Agent Orange case, like *Allen,* began five years ago and involved thousands of veterans who claimed billions of dollars in damages due to the injuries that occurred from toxic chemical exposure during the Vietnam War. Judge Weinstein, concerned about the military exceptions in the FTCA to tort claims, "helped spur the settlement by innovative rulings narrowing the defense of the government or the herbicide's manufacturers." In a number of rulings, he held that the *Feres* doctrine did not apply to wives and children, and that the government could be made "to pay a share of any damage for miscarriages and birth defects attributable to genetic damage caused by Agent Orange. [He] also stretched the law to give the veterans a shot at the difficult task of establishing that Agent Orange was not the cause of their injuries." But he warned that appeals court might not follow him *so far beyond the mainstream of the law.* See Stuart Taylor, Jr., "Why the Army Has An Edge in Court," *The New York Times,* May 13, 1984. See also Fred Wilcox, *Waiting for an Army to Die* (New York: Vintage Books, 1983).

101. Stephen Soble, "A Proposal for the Administrative Compensation of Victims of Toxic Substance Pollution," *Harvard Journal on Legislation,* 14 (1977), p. 709.

102. Tilevitz, "Judicial Attitudes Toward Legal and Scientific Proof of Cancer Causation," *Columbia Journal of Environmental Law,* 3 (1977), p. 345. Tilevitz concluded by stating that the problem could be eliminated if the federal judges "would accept as probative testimony that a substance 'could have caused' a cancer, and realize that in scientific terminology, such testimony is more than just speculative," p. 379.

103. Schaffer report, *Compensation,* pp. 37–41.

104. Favish, "Radiation Injury and the Atomic Vet," *Hastings Law Journal,* 32 (March, 1981). Favish argues that the burden of proof in these cases should shift to the AEC and the government lawyers because (1) they knew of radiation hazards, (2) they were responsible for the inadequate dosimetry and, (3) the downwinders and the veterans were involuntarily used by the AEC. In sum, there was an "evidentiary void," which was the direct and forseeable result of the government's very unreasonable conduct during the testing program, pp. 972–973ff. However, even with the shift in burden of proof, the government can probably disprove the causal link.

105. *Ibid.,* p. 974.
106. Schaffer Report, *Compensation,* p. 41.
107. Janet Gordon, interview with author, May 14, 1984.
108. Daniel Schorr, *National Public Radio,* May 14, 1984.
109. Jenkins, in *Allen,* 1981, p. 9.
110. Editorial, "Ruling Underlines Dire Need For New Damage Guidelines," *Salt Lake Tribune,* May 12, 1984.

CHAPTER EIGHT

1. See *National Journal,* September 24, 1983, p. 1931.
2. Titus, *Governmental Responsibility,* p. 278.
3. *Ibid.*
4. *National Journal,* September 24, 1983, p. 1931.
5. Larry Gersten, *Making Public Policy: From Conflict to Resolution.* (Glenview, Illinois: Scott, Foresman, and Co., 1983), p. 24. See also Schaffer report, *Compensation,* pp. 53–54.
6. Howard Ball, Dale Krane, and Tom Lauth, *Compromised Compliance: Implementation of the 1965 Voting Rights Act* (Westport, Connecticut: Greenwood Press, 1982), p. 110.
7. Gersten, *Making Public Policy,* p. 23. A story in the May 25, 1984 *Deseret News* indicates the continuing dilemma of cancer and radiation. The UPI story, "Study Says Rate of Cancer Is High Near N-Laboratory," stated that "a high cancer rate has been noted among residents downwind from the Idaho National Engineering Laboratory, a nuclear research and testing center on the Snake River Plain."
8. See generally, E. E. Schattschneider, *The Semi-Sovereign People, A Realist's View of Democracy in America* (New York: Holt, Rinehart, and Winston, 1960).
9. See S. 1480, "Superfund" legislation, introduced in 1980 by Senator Edmund S. Muskie (D-Maine), and Chapter 5, "Implementing EO 12,291: The Politics of the Toxic Chemical Labeling Standard," in Howard Ball, *Controlling Regulatory Sprawl* (Westport, Connecticut, Greenwood Press, 1984).
10. See, for example 98th Congress: S.921, *Radiogenic Cancer Compensation Act of 1983;* S.1483, *Radiation Exposure Compensation Act of 1981;* S.228, *Federal Radiation Protection Management Act of 1982;* S.1205, *Environmental Research, Development, and Demonstration Act of 1982;* S.26, *Comprehensive Vietnam and Post-Vietnam Era Veterans' Readjustment, Recruitment, and Retention Assistance Act of 1981;* HR.5152, 5153, 4012, 1763, 7285, 1564, 1733, 1769, 2229, 673, *Modification in VA Legislation;* HR.3499, *Veterans' Health Care Act of 1981;* 97th Congress: HR 1022, VA legislation; HR.4544, 1914, 1943, 1022, 1961, 2903; HR.3909, *Atomic Veterans' Relief Act;* S.11, *Veterans' Programs Improvement Act of 1983;* S.991, *VA legislation;* S.2166, *Indian Health Care Amendments of 1983;* HR.1961, *Vietnam Veterans' Agent Orange Relief Act;* S.1651, *Veterans' Dioxin and Radiation Exposure Compensation Standards Act of 1983;* S.2166, *Indian Health Advisory Board.*
11. Titus, *Governmental Responsibility,* p. 278.
12. Jeffrey Trauberman, "Compensating Victims of Toxic Substances Pollution: An Analysis of Existing Federal Statutes," *Harvard Environmental Law Review,* 5 (1981), p. 24–25.
13. *Ibid.*
14. *1978 Congressional Quarterly Almanac,* pp. 266–267.
15. Dr. Ronald Preston, Majority Staff, Senate Labor and Human Resources Committee, interview with Research Associate Kenneth Verdoia, Washington, D.C., May 17, 1984. Claims under the Black Lung Benefits Act have far exceeded earlier legislative estimates. From 1969 to 1981, benefit expenditures under the Act stand at almost $10 billion. See the Hearings Before House Subcommittee on Oversight, Committee on Ways

and Means, *Black Lung Compensation Legislation*, 97th Congress, 1st session, 1981.

16. *Ibid.*

17. *Radiation Exposure Compensation Act of 1981*, pt. I, October 27, 1981, p. 36.

18. The Corporation has sued for bankruptcy in the U.S. Bankruptcy Court for the Southern District, New York.

19. Preston, interview with Verdoia, May 17, 1984.

20. Elizabeth Wehr, "Victims of Radiation, Other Environment-Related Illness Seeking Help from Congress," *Congressional Quarterly*, February 23 (1980), p. 552.

21. Preston, interview with Verdoia, May 17, 1984.

22. Trauberman, "Compensating Victims," p. 1ff.

23. *Ibid.*, p. 6.

24. *Ibid.*

25. *Ibid.*, pp. 8–13ff.

26. *Ibid.*, p. 12.

27. *Ibid.*, p. 28.

28. Ronald Brownstein, "Asbestos Litigation, a Legal Nightmare that Congress is Being Asked to End," *National Journal*, September 24 (1983), p. 1942.

29. *Ibid.*, p. 1943.

30. *Ibid.*

31. Wehr, "Victims of Radiation," p. 553.

32. *Ibid.*

33. Gersten, *Making Public Policy*, p. 61.

34. *Ibid.*, p. 62.

35. Timothy B. Clark, "Stiff Tax Hikes Will Be Key To Future Efforts To Close Budget Deficit," *National Journal*, April 21 (1984), p. 753.

36. Wehr, "Victims of Radiation," p. 552.

37. Preston, interview with Verdoia, May 17, 1984.

38. Senior minority staff person, Labor and Human Resources Committee, telephone interview with author, May 22, 1984.

39. *Ibid.*

40. Wehr, "Victims of Radiation," p. 554.

41. *Ibid.* These groups supported "Superfund" legislation to clean up toxic wastes spilled or dumped into the environment. See Kathy Koch, "Compromise Superfund Proposal Cleared," *Congressional Quarterly*, December 6, 1980, p. 3509.

42. Janet Gordon, interview with author, Salt Lake City, Utah, May 15, 1984.

43. Preston, interview wtih Verdoia, May 17, 1984.

44. *Ibid.*

45. *Ibid.*

46. *Ibid.*

47. *Ibid.*

48. S.1865, *Radiation Exposure Compensation Act of 1979*, 96th Congress.

49. *Ibid.*

50. *Ibid.*

51. Schaffer, *Compensation*, pp. 42–43.

52. *Radiation Exposure Compensation Act of 1979*, 1980 Hearings, p. 51.

53. *Ibid.*, p. 52.

54. Schaffer, *Compensation*, p. 45.

55. *Ibid.*, pp. 45–46.

56. *Ibid.*, pp. 46–47.

57. *Radiation Exposure Compensation Act of 1979*, 1980 Hearings, p. 53.

58. *Ibid.*, p. 61.

59. *Ibid.*, p. 67.

60. *Ibid.*, p. 71.

61. *Ibid.*, p. 54.

62. *Ibid.*

63. Michael Barone and Grant Ujifusa, *The Almanac of American Politics,* Washington, D.C.: *National Journal,* 1984, pp. xxxiii–xxxiv.

64. *Ibid.,* p. 1185.

65. *Ibid.*

66. *Ibid.*

67. Preston, interview with Verdoia, May 17, 1984, Washington, D.C.

68. *Ibid.*

69. S.1483, *Radiation Exposure Compensation Act of 1981,* 97th Congress.

70. *Ibid.*

71. Testimony of General Griffith, *Radiation Exposure Compensation Act of 1981,* Hearings, pt. I, October 27, 1981, p. 49. The following bureaucrats were present at the hearing: Dr. Edward N. Brandt, Assistant Secretary, Health and Human Services and Director, Interagency Radiation Research Committee; accompanied by Dr. Charles Lowe, Special Assistant to the Director, National Institutes of Health, Technical Consultant; Dr. Edwin T. Still, Assistant to the Director, Defence Nuclear Agency, Department of Defense; Dr. Charles W. Edington, Deputy Director, Office of Health and Environmental Research, Department of Energy; Dr. Oddvar Nygaard, Special Assistant to the Director, National Cancer Institute, National Institutes of Health, Department of Health and Human Services; Dr. Matt Kinnard, Medical Officer, Veterans Administration; Gerald Rausa, Health Research Administrator, Environmental Protection Agency; Lt. Gen. Harry A. Griffith, U.S. Army, Director, Defense Nuclear Agency, Department of Defense, accompanied by Mahlon E. Gates, Manager, Nevada Operations Office, Department of Energy; Dr. Lynn R. Anspaugh, Section Leader, Lawrence Livermore National Laboratory; Bruce W. Church, Director, Health Physics Division, Nevada Operations Office, Department of Energy; Dr. Edwin T. Still, Assistant to the Director, Defense Nuclear Agency; Robert L. Brittigan, General Counsel, Defense Nuclear Agency.

72. *Ibid.,* p. 35.

73. *Ibid.,* p. 39.

74. *Ibid.,* pp. 46–47.

75. *Ibid.,* p. 56.

76. *Ibid.,* p. 74.

77. *Ibid.,* p. 261.

78. *Ibid.,* p. 262.

79. *Ibid.,* p. 263.

80. *Ibid.,* p. 264.

81. *Ibid.,* p. 267.

82. *Ibid.,* pp. 267–271.

83. See press release, "Administration 'Open' To Plan Compensating Fallout Victims," March 12, 1982, Senator Orrin Hatch, Washington, D.C.

84. See press release, "Human Resources Committee Approves Atomic Bomb Fallout Compensation Bill," April 20, 1982, Senator Orrin Hatch, Washington, D.C., p. 1.

85. *Radiation Exposure Compensation Act of 1981,* Hearings pt. 2, April 8, 1982, p. 16.

86. Press release, "Human Resources Committee Approves Bill," Senator Hatch, p. 1.

87. Titus, *Governmental Responsibility,* p. 288.

88. *Ibid.*

89. Elisabeth Wehr, "Bill to Help Radiation Victims Collect Damages Bogs Down," *Congressional Quarterly,* July 3 (1982), p. 1588.

90. *Ibid.*

91. Senior minority staff person, Senate Labor and Health Resources Committee, interview with author, May 22, 1984.

92. *Ibid.*

93. Preston, interview with Verdoia, May 17, 1984.

94. *Ibid.*

95. S.921, *Radiogenic Cancer Compensation Act of 1983,* 98th Congress.

96. Preston, interview with Verdoia, May 17, 1984.

97. Governor Scott Matheson, interview with author, Salt Lake City, Utah, May 10, 1984.

98. Senior minority staff person, interview with author, May 22, 1984.

99. *Ibid.*

100. *Ibid.*

101. Hatch's testimony before Senate Committee on Energy and Natural Resources, May 24, 1984.

102. Preston, interview with Verdoia, May 17, 1984.

103. Thomas H. Gorey, "Hatch Suggests Funds for N-Test Residents," *Salt Lake Tribune,* May 26, 1984.

104. Hatch's testimony before Senate Committee on Energy and Natural Resources, May 24, 1984.

105. Gorey, "Hatch Suggests Funds," *Salt Lake Tribune.*

106. *Ibid.*

107. Senior minority staff person, interview with author, May 22, 1984.

108. Preston, interview with Verdoia, May 17, 1984.

109. See *Deseret News,* May 13, 1984, and *Salt Lake Tribune,* May 12, 1984.

110. *Deseret News,* "Fallout Verdict" editorial, May 13, 1984.

111. Gordon E. White, "Court Decision On A-Fallout Clears The Way For Hatch Bill," *Deseret News,* May 12, 1984.

112. "Fallout Verdict," *Deseret News.*

113. Quoted in *Salt Lake Tribune,* May 11, 1984.

114. Governor Scott Matheson, interview with author, May 10, 1984.

115. Gordon E. White, "Marriott Will Renew Effort To Compensate Victims," *Deseret News,* May 13, 1984.

116. In late May, 1984, efforts of Utah Governor Scott Matheson and Congressman Dan Marriot to move a pile of irradiated tailings from the middle of Salt Lake County were set aside in a House Report that neglected to inform the outgoing congressman of its negative recommendation. He found out about the decision not to fund the clean-up when he read the report on the 1985 energy and water appropriations bill. See "Senate Panel Rejects Tailings' Removal," *Salt Lake Tribune,* May 18, 1984.

117. Senator Hatch's testimony on May 24, 1984 suggested that Jenkins' "effort to provide justice within the law was a bit too creative." See his press release dated May 24, 1984. p. 2.

118. White, "Court Decision on A-Fallout," *Deseret News.*

119. See Robert Rothman, "House Approves Agent Orange Disability Pay," *Congressional Quarterly,* February 4 (1984), p. 229; "Winning Peace With Honor," *Time,* May 21, 1984, p. 40.

CHAPTER NINE

1. Judge Charles D. Breitel, *Superfund* Study Group, Minority Report, U.S. Congress, 98th Congress, 1983, p. 2.

2. Lynton K. Caldwell, *Science and the National Environmental Protection Act* (University of Alabama: University of Alabama Press, 1982), p. 37.

3. *Ibid.,* p. 27.

4. *Ibid.,* p. 32.

5. Sissela Bok, *Lying: Moral Choice in Public and Private Life* (New York: Pantheon Books, 1978), p. 19.

6. Quoted in Caldwell, *Science and Environmental Protection,* p. 27.

7. *Ibid.,* p. 37.

8. Breitel, *Superfund* Report, 1983, p. 2.

9. According to John O'Connor, Chairman of the National Campaign Against Toxic Hazards, toxic wastes illegally dumped have accounted for the national increase of cancer. He claimed that "up to 90% of all cancers may be caused by exposure to poisonous chemicals." See Ernest H. Linford, "Another Case Proves Peril of Pollution," *Salt Lake Tribune,* May 16, 1984. Samuel S. Epstein, in his *Politics of Cancer,* (San Francisco: Sierra Books, 1978) has written that "most cancer (70 to 90%) is environmental in origin and is therefore preventable," p. 23.

10. George Freeman, Jr., "The Superfund Section 301(e) Study Group Recommendations and Related Pending Proposals," paper presented at the National Conference on Environmental Injury Compensation, Washington, D.C., March, 1981, pp. 25–26. With regard to the atomic veterans, the Veterans Administration changed VA regulations on April 6, 1983, to provide free medical treatment to veterans who took part in the open-air tests and who subsequently developed illnesses, in addition to leukemia and cancer, that might have been the result of exposure to radiation. The new guidelines, published later in the *Federal Register,* were made more liberal "to allow treatment of all conditions except for those which are known to have causes other than radiation exposure," said a VA spokesperson before a Senate Veterans Affairs Committee hearing in Washington, D.C. See "U.S. Expands Care of Atom Veterans," *The New York Times,* April 7, 1983.

11. Freeman, "Superfund Study Group Recommendations," 1984, p. 26.

12. *Ibid.,* p. 58.

13. Judge Breitel, and others involved with the federal government, has cautioned about the adverse impact of unfeasible compensation packages: "The society able to provide adequate remedies on the scale required must also be economically productive and its enterprises capable of surviving economically in a competitive world no longer confined to the opposite sides of the North Atlantic Ocean," *Superfund* Report, 1983, p. 8.

14. Judge Bruce S. Jenkins, *Allen v. United States,* Civil No. C-79-0515-J, 1984, at p. 317.

15. See Jenkins' *Allen* opinion, pp. 205–236, and Favish, *Atomic Veterans,* pp. 938–942.

16. Favish, *Atomic Veterans,* p. 938.

17. See *Allen v. United States,* at pp. 281–286, for Jenkins' review of the Dirty Harry cover-up.

18. E. H. Haskell, P. L. Kaipa, and M. E. Wrenn, "The Use of Thermoluminescence Analysis for Atomic Bomb Dosimetry: Estimating and Minimizing Total Error," in *Reassessment of Atomic Bomb Radiation Dosimetry in Hiroshima and Nagasaki,* Second United States–Japan Workshop, Radiation Effects Research Foundation, 1983, p. 38.

19. Epstein, *Politics of Cancer,* p. 316.

20. *Ibid.,* p. 317.

21. *Ibid.*

22. Robert Gilmour, "Congressional Oversight," *Bureaucrat,* 10 (Fall, 1981), p. 18.

23. *Allen,* at p. 6.

24. *Ibid.,* at p. 9.

25. *Ibid.,* pp. 8–9.

26. *Ibid.,* at pp. 195–196.

27. Breitel, *Superfund* Report, 1983, p. 8.

POSTSCRIPT

1. *Alice P. Broudy v United States, et al.,* US District Court, C.D. CA, Civil No. 79-2626-LEW, *Memorandum In Support of Defendant's Motion For Summary Judgment,* November 12, 1985, Attachment 4, "Clearance of "Discussion on Product Contamination after underwater bomb detonation" during part of informal talks at Bikini," 9 October 1946.

The *Broudy* litigation was the effort of the widow of a Major in the U.S. Marines to recover, in a tort action under the FTCA, damages from the government due to her husband's death from cancer caused by his exposure to radioactive fallout while he participated in Nevada Test Site field activities close to ground zero in 1957. Her claim for damages was based on the premise that the government was negligent with her husband *after* he was discharged from the military and that the injury fell into the post-discharge tort category. She made that claim because of the existence of the *Feres* doctrine, in which the U.S. Supreme Court held that the government was not liable, under the FTCA exception, for injuries to military personnel when injuries, such as cancer from exposure to radiated fallout, arose out of military activities. For the government to win the suit, it had to show the court that it knew of the hazards of radioactivity prior to Broudy's discharge and that failure to warn him was a policy decision of the government that was covered by the discretionary function exception in the FTCA. If the DOJ could show this official knowledge of the hazard, *Feres* bars suit for any failure to warn after discharge. As a consequence, the government produced documents to show, clearly, that as early as 1945–46, the government *knew* that radioactive fallout was "the most toxic metal known," and that it would cause "cancer, leukemia, birth defects, histopathological changes, genetic damage, and biological and enzymatic disturbances" *Defendant's Brief,* p. 10. The federal judge hearing the arguments for and against summary dismissal evidently believed that the government knew of the dangers and, on November 25, 1985, ruled for the defendant and summarily dismissed the suit.

2. *Ibid.,* pp. 8–9.

3. *Ibid.,* Attachment 4, pp. 1–2.

4. *Ibid.,* Attachment 3, Classified *SECRET.* "Lecture to Radiological Safety Personnel," October 7, 1946, pp. 21–22.

5. *Ibid.,* p. 13.

6. See Sissela Bok, *Lying: Moral Choice in Public and Private Life,* New York: Vintage Books, 1978. Officials may lie for "noble purposes, [but] if we assume the perspective of the deceived—those who experience the consequences of governmental deception—such arguments are not persuasive," p. 178. See also Judge S. Christensen's opinion in *Bulloch, II.*

APPENDIX A

1. Uhl and Ensign, *GI Guinea Pigs,* pp. 4–6. The doctors' conclusions "correlated the sheep deaths to exposure to nuclear radioactive fallout," p. 26.

2. AEC, *Radiation Protection Handbook,* No. 59, 1951.

3. *Health Effects of Low Level Radiation,* 1979 Hearings, pp. 163–165.

4. *Ibid.,* pp. 286, 288. Johnson notes that the fetal lambs had "received doses of 20,000 to 40,000 rads to the thyroid gland, and sheep had received 1,500 to 6,000 rads to the gastrointestinal tract, where the external doses were estimated to be only 4 rads. "Cancer Incidence," p. 231.

5. *Radiation Exposure,* p. vii.

6. *Health Effects of Low Level Radiation,* 1979 Hearings, p. 98.

7. *Bulloch v. United States,* 95 F Rules Dec 123 (1982), p. 125.

8. *Bulloch v. United States,* 145 F Supp 824 (1956), pp. 829–830.

9. *Id.,* at 830.

10. Wrote Judge Christensen, "I can't help but believe that those gentlemen were conscious of the implications; [untruths about the sheep deaths would] jeopardize every one of us, because the general pattern could extend to humans." *Health Effects,* p. 18. *Id.* at 834.

11. *Ibid.*

12. See their testimony in *Health Effects of Low Level Radiation,* 1979 Hearings, Vol. I, pp. 69ff., esp. p. 98, where Libassi states that the dose levels in the animals "were 1,000 times the permissible count for human beings in the thyroid, and 50% higher than the permissible levels in the bone marrow."

13. *Health Effects of Low Level Radiation,* 1979 Hearings, p. 18.

14. *Bulloch v. United States,* 1982.

15. *Ibid.,* at 131–142.

16. *Id.,* at 133.

17. *Id.,* at 135.

18. *Id.,* at 136.

19. *Id.,* at 138.

20. *Id.,* at 143.

21. *Id.,* at 144.

22. *Id.,* at 143.

23. *Id.*

24. Letter, Judge A. Sherman Christensen, to author, January 22, 1985.

25. *Ibid.*

26. *Ibid.*

27. *Bulloch v. United States,* 82-2245, 82-2352 (Civil No. C-81, 0123C), November 23, 1983, p. 17 (Slip Opinion).

APPENDIX B

1. "The fact is that no one knows precisely the effects of low-level radiation over long periods of time: As a result, the scientific community is polarized in a raging debate." Michio Kaku and Jennifer Trainer, eds., *Nuclear Power: Both Sides* (New York: W. W. Norton, 1982), p. 28.

2. Fabrikant, "Epidemiological Studies," p. 461.

3. See, for example, Seymour Jablon, "Radiation," in Joseph Franmeni, Jr., ed., *Persons at High Risk of Cancer,* pp. 151–165.

4. A. Bertrand Brill, ed., *Low-Level Radiation Effects: A Fact Book,* (New York: Society of Nuclear Medicine, Inc., 1982), ch. 2.

5. Rosalyn S. Yalow, "Reappraisal of Potential Risks Associated with Low-Level Radiation," *Annals of the New York Academy of Sciences* (November, 1981), p. 49.

6. Gofman, *Radiation,* p. 368.

7. Yalow, "Reappraisal of Potential Risks," p. 37. Gofman, a critic of Yalow, also talks of proportionality in the linear model, asking "If 10 rads cause 1,000 cancers in a large population, will 1 rad cause 100 cancers, and will 0.1 rad cause 10 cancers; and will 0.01 Rad cause 1 cancer?", *Radiation,* p. 56.

8. Gofman, *Radiation,* p. 47.

9. *Ibid.*

10. Jablon, in *Persons at High Risk,* p. 153.

11. Brill, *Low-Level Radiation Effects,* pp. 2–9.

12. Schaffer Report, *Compensation,* p. 18. See also Hardee, "Real and Perceived Risks," p. 382.

13. Quoted on KUTV, "Downwind," December 17, 1982. See also Roberts, "Genetic Epidemiology," p. 67.

Bibliography

Books

Allen, James B. and Glen M. Leonard. *The Story of the Latter Day Saints*. Salt Lake: Deseret Book Co., 1976.

Arrington, Leonard J. and Davis Bitton. *The Mormon Experience: A History of the Latter-Day Saints*. New York: Alfred A. Knopf, 1979.

Ball, Howard. *Controlling Regulatory Sprawl*. Westport, Connecticut: Greenwood Press, 1984.

Ball, Howard, Dale Krane, and Thomas P. Lauth. *Compromised Compliance: Implementation of the 1965 Voting Rights Act*. Westport, Connecticut: Greenwood Press, 1982.

Barone, Michael and Grant Ujifusa. *The Almanac of American Politics*. Washington, D.C.: National Journal, 1984.

Bok, Sissela. *Lying: Moral Choice in Public and Private Life*. New York: Pantheon Books, 1978.

Boorstein, Daniel J. *The Americans: The Democratic Experience*. New York: Random House, 1973.

Brill, Bertrand, ed. *Low-Level Radiation Effects: A Fact Book*. New York: Society of Nuclear Medicine, 1982.

Caldwell, Lynton K. *Science and the National Environmental Protection Act*. University of Alabama: University of Alabama Press, 1982.

Califano, Joseph A., Jr. *Governing America: An Insider's Report from the White House and the Cabinet*. New York: Simon and Shuster, 1981.

Clark, John G., et al. *Three Generations in Twentieth Century America: Family, Community and Nation*. Homewood, Illinois: Dorsey Press, 1977.

Davis, David D. *Energy Politics*. New York: St. Martin's Press, 1974.

Donovan, Robert J. *The Tumultuous Years: The Presidency of Harry S Truman, 1949–1953*. New York: Norton, (1967) 1982.

Epstein, Samuel S. *Politics of Cancer*. San Francisco: Sierra Club Books, 1978.

Fisher, Albert. *Geography of Utah*. Salt Lake: University of Utah Press, 1980.

Frankl, Victor. *Man's Search for Meaning*. New York: Washington Square Press, 1963.

Freeman, Leslie J. *Nuclear Witnesses*. New York: Norton, 1982.

Fuller, John G. *The Day We Bombed Utah*. New York: New American Library, 1984.

Funston, Richard. *Constitutional Counter-Revolution? The Warren Court and The Burger Court: Judicial Policy-Making in Modern America*. New York: John Wiley and Sons, 1977.

Gerber, Judy and Martin E. P. Seligman, eds. *Helplessness: Theory and Applications*. New York: Academic Press, 1980.

Gersten, Larry. *Making Public Policy: From Conflict to Resolution*. Glenview, Illinois: Scott, Foresman, and Co., 1983.

Gofman, John. *Radiation and Human Health*. San Francisco: Sierra Club Books, 1981.

Gottlieb, Robert and Peter Wiley. *America's Saints: The Rise of Mormon Power*. New York: G. P. Putnam's Sons, 1984.

Goulden, Joseph C. *The Best Years, 1945–1950*. New York: Atheneum, 1976.

Grodzins, Morton and Eugene Rabinowich. *The Atomic Age: Scientists in National and World Affairs*. New York: Basic Books, 1963.

Groves, Leslie M. *Now It Can Be Told: The Story of the Manhattan Project*. New York: DeCapo Press, 1962.

Herkin, Gregg. *The Winning Weapon: The Atomic Bomb in the Cold War, 1945–1950*. New York: Knopf, 1980.

Hewlett, Richard G. and Oscar E. Anderson, Jr. *The New World, 1939–1946. Vol. 1, A History of the U.S. Atomic Energy Commission*. University Park, Pennsylvania: Pennsylvania State University Press, 1962.

Hewlett, Richard G. and Francis Duncan. *Atomic Shield, 1947–1952. Vol. 2, A History of the U.S. Atomic Energy Commission*. University Park, Pennsylvania: Pennsylvania State University Press, 1969.

Ishikawa, Eisei and David L. Swain. *Hiroshima and Nagasaki: The Physical, Medical, and Social Effects of the Atomic Bombings*. New York: Basic Books, 1979.

Kaku, Michio and Jennifer Trainer, eds. *Nuclear Power: Both Sides*. New York: W. W. Norton, 1982.

Kunetka, James W. *City of Fire: Los Alamos and the Atomic Age, 1943–1945*. Albuquerque: University of New Mexico Press, 1979.

LeBaron, Wayne D. *The Reluctant Survivors*. Salt Lake City: Dream Garden Press, 1985.

Mayfield, James B. *The First 80 Years: Utah Voting Behavior in Perspective*. Unpublished. Salt Lake: 1980.

Metzger, H. Peter. *The Atomic Establishment*. New York: Simon and Schuster, 1972.

Nelson, Lowry. *The Mormon Village: A Pattern and Technique of Land Settlement*. Salt Lake: University of Utah Press, 1952.

Nelson, Elroy. *Utah's Economic Patterns*. Salt Lake: University of Utah Press, 1956.

Poll, Richard, ed. *Utah's History*. Provo, Utah: Brigham Young University Press, 1978.

Prosser, William Lloyd. *Law of Torts*. St. Paul: West Publishing Co., 1971.

Saffer, Thomas H. and Orville E. Kelly. *Countdown Zero*. New York: Penguin Books, 1982.

Schattschneider, E. E. *The Semi-Sovereign People, A Realist's View of Democracy in America*. New York: Holt, Rinehart, and Winston, 1960.

Smyth, Henry DeWolf. *Atomic Energy for Military Purposes: The Official Report on the Development of the Atomic Bomb Under the Auspices of the U.S. Government, 1940–1945*. Princeton, New Jersey: Princeton University Press, 1945.

Susser, Mervyn. *Causal Thinking in the Health Sciences*. New York: Oxford University Press, 1973.

Tyler, Abell, ed. *Drew Pearson Diaries, 1949–1959*. New York: Holt, Rinehart, and Winston, 1974.

Uhl, Michael and Tod Ensign. *G. I. Guinea Pigs*. Chicago: Westview, 1980.

Watson, James D. *The Double Helix*. New York: Atheneum, 1969.

White, Theodore. *Breach of Faith: The Fall of Richard Nixon*. New York: Atheneum, 1975.

Wilcox, Fred A. *Waiting For an Army to Die*. New York: Vintage Books, 1983.

Wilkinson, J. Harvie, III. *From Brown to Bakke*. New York: Oxford University Press, 1980.

Articles

Aberbach, Joel D. and Bert A. Rockman. "Clashing Beliefs Within the Executive Branch," *American Political Science Review* 70 (June, 1976).

Ball, Howard. "The Nation State: Immorality and Violence," *Sunstone* 7, no. 6 (November/December, 1982).

————. "The Problems and Prospects of Fashioning a Remedy for Radiation Injury Plaintiffs in Federal District Court: Examining *Allen v. United States, 1984,*" *University of Utah Law Review* no. 2 (Summer, 1985).

Baum, Andrew. "Coping with Chronic Stress at Three-Mile Island," *Health Psychology* 2 (1983).

Baum, Andrew, Raymond Fleming, and Jerome Singer. "Coping with Victimization by Technological Disaster," *Journal of Social Issues* 39, no. 2 (1983).

Beck, Harold L. and Philip W. Krey. "Radiation Exposures in Utah from the Nevada Nuclear Tests," *Science* 220 (April 1, 1983).

Bond, V. P. and L. D. Hamilton. "Leukemia in the Nevada Smoky Bomb Test," *Journal of the American Medical Association* 244, no. 14 (October 3, 1980).

Brownstein, Ronald. "Asbestos Litigation, a Legal Nightmare that Congress Is Being Asked To End," *National Journal* (September 24, 1983).

Caldwell, Glyn G., Delle B. Kelly, and Clark W. Heath, Jr. "Leukemia Among Participants in Military Maneuvers at a Nuclear Bomb Test," *Journal of the American Medical Association* 244, no. 14 (October 3, 1980).

Caldwell, Glyn G., et al. "Mortality and Cancer Frequency Among Nuclear Test (Smoky) Participants, 1957–1979," *Journal of the American Medical Association* 250, no. 5 (August 5, 1983).

Clark, Timothy B. "Stiff Tax Hikes Will Be Key to Future Efforts To Close Budget Deficit," *National Journal* (April 21, 1984).

Clayton, James L. "Contemporary Economic Development." In *Utah's History* edited by Richard Poll. Provo, Utah: Brigham Young University Press, 1978.

Davis, Laura, Andrew Baum, and Daniel Collins. "Stress and Control-Related Problems at Three-Mile Island," *Journal of Applied Social Psychology* 12, no. 5 (1982).

Dean, Gordon. "The Impact of the Atom on Law," *University of Pittsburgh Law Review* 128 (1951).

Delgado, Richard. "Beyond *Sindell*: Relaxation of Cause-in-Fact Rules for Indeterminate Plaintiffs," *California Law Review* 70 (1982).

Doyle, Christine M. "Government Liability for Nuclear Testing Under FTCA," *University of California, Davis, Law Review* 15 (1982).

England, Eugene. "Can Nations Love Their Enemies: An LDS Theology of Peace," *Sunstone* no. 6 (November/December, 1982).

Estep, Samuel D. "Radiation Injuries and Statistics," *Michigan Law Review* 59 (1960).

Fabrikant, Jacob I. "Epidemiological Studies on Radiation Carcinogenesis in Human Populations Following Acute Exposure: Nuclear Explosions and Medical Radiation," *Yale Journal of Biology and Medicine* 54 (1981).

Favish, Allan. "Radiation Injury and the Atomic Veteran: Shifting the Burden of Proof on Factual Causation," *Hastings Law Journal* 32 (March 1981).

Fleming, Raymond, et al. "Mediating Influences of Social Support on Stress at Three-Mile Island," *Journal of Human Stress* (September, 1982).

Folley, J. H., Wayne Borges, and T. Yamawaki, "Incidence of Leukemia in Survivors of the A-Bomb in Hiroshima and Nagasaki," *American Journal of Medicine* 13 (1952).

Franco, Adolpho. "*Wollman v. Gross*: Statute of Limitations and the FTCA," *Creighton Law Review* 15 (1982).

Fuller, John G. "The Day We Bombed Utah," *Omni* (October, 1982).

Furth, J. "Recent Studies on the Etiology and Nature of Leukemia," *Blood* 6 (1951).

Furth, J. and O. B. Furth. "Neoplastic Diseases Produced in Mice by General Irradiation with X-Rays," part 1, *American Journal of Cancer* 28 (1936).

Gilmour, Robert. "Congressional Oversight," *Bureaucrat* 10 (Fall, 1981).

Gofman, John, et al. "Low Dose Radioactivity, Chromosomes and Cancer," IEEE (October, 1969).

Goldman, Sheldon. "A Profile of Carter's Judicial Nominees," *Judicature* 62 (November, 1978).

Green, Leon. "The Duty Problem in Negligence Cases," *Columbia Law Review* 28 (1928, 1929).

Hendee, William R. "Real and Perceived Risks of Medical Radiation Exposure," *Western Journal of Medicine* 138 (March, 1983).

Hill, Marvin S. "The Rise of the Mormon Kingdom of God." In *Utah's History,* edited by Richard Poll. Provo, Utah: Brigham Young University Press, 1978.

Jablon, Seymour. "Radiation." In *Persons at High Risk of Cancer: An Approach to Cancer Etiology and Control,* edited by Joseph Fraumeni, Jr. Conference Proceedings, Key Biscayne, Florida, December 10-12, 1974.

Jenkins, Bruce S. "Communication in the Courtroom: The Lawyer as an Educator," *American Bankruptcy Law Journal* 48 (1974).

————. "Remarks Before Joint Meeting," *Utah Bar Journal* 10 (January, 1982).

Johnson, Carl J. "Cancer Incidence in an Area of Radioactive Fallout Downwind from the Nevada Test Site," *Journal of the American Medical Association* 251 (January 13, 1984).

Johnson, Frank M., Jr. "Federal Judges: Policymakers," *Utah Forum* 1 (1981).

————. "The Role of the Federal Courts in Institutional Litigation," *Alabama Law Review* 32 (1981).

Kato, Hiroo, et al. "Studies of the Mortality of A-Bomb Survivors," *Radiation Research* 90, 91 (1982).

Keeton, R. E. "Creative Continuity in the Law of Torts," *Harvard Law Review* 75 (1962).

Koch, Kathy. "Compromise Superfund Proposal Cleared," *Congressional Quarterly* (December 6, 1980).

Land, Charles. Critique of "The Hazards of Fallout or of Epidemiological Research?" by Dr. Joseph Lyon, *New England Journal of Medicine* 300, no. 8 (February 22, 1979).

Land, Charles, et al. "Childhood Leukemias and Fallout from the Nevada Nuclear Tests," *Science* 223 (January 13, 1984).

Lange, Robert D., William C. Moloney, and T. Yamawaki. "Leukemia in A-Bomb Survivors, In General Observations," *Blood* 9 (1954).

Larsen, Gustive L. "Government, Politics, and Conflict" and "The Crusade and the Manifesto." In *Utah's History,* edited by Richard Poll. Provo, Utah: Brigham Young University Press, 1978.

Lewis, E. B. "Leukemia and Ionizing Radiation," *Science* 125 (1957).

————. "The Biological Effects of Atomic Radiation," *National Archives of Science, National Research Council* (1956).

————. "The Hazards to Man of Nuclear and Allied Radiations," *British Medical Research Council* (1956).

Lightman, Alan P. "To Cleave an Atom," *Science* 5 (November, 1984).

Lisco, H., M. P. Finkel, and A. M. Brues, "Carcinogenic Properties of Radioactive Fission Products and of Plutonium," *Radiology* 49 (1947).

Lyon, Joseph L. and John W. Gardner, "Radiation Exposure and Cancer," *Journal of the American Medical Association* 246, no. 19 (November 13, 1981.

Lyon, Joseph L., et al. "Childhood Leukemias Associated with Fallout from Nuclear Testing," *New England Journal of Medicine* 300 (February 22, 1979).

March, Herman C. "Leukemia in Radiologists in a 20-Year Period," *American Journal of Medical Science* 220 (1950).

Moore, Charles. "Radiation and Preconception Injuries," *Southwestern Law Review* 28 (1974).

Murphy, Eileen M. "Atomic Energy and the Law," *Vanderbilt Law Review* 12 (1958).

Niehoff, Richard O. "Organization and Administration of the United States Atomic Energy Commission," *Public Administration Review* 8 (Spring, 1948).

Peterson, F. Ross. "Utah Politics Since 1945." In *Utah's History,* edited by Richard Poll. Provo, Utah: Brigham Young University Press, 1978.

Rallison, Marvin, et al. "Thyroid Disease in Children," *American Journal of Medicine* 56 (April, 1974).

Raloff, Janet. "Compensating Radiation Victims," *Science News* 124 (November 19, 1983).

Remsberg, Bonnie. "Beneath the Cloud," *Family Circle* (November 15, 1983).

Rothman, Robert. "House Approves Agent Orange Disability Pay," *Congressional Quarterly* (February 1, 1981).

————. "Winning Peace With Honor," *Time* (May 21, 1984).

Sholl, Debra. "The Nevada Proving Grounds: An Asylum for Sovereign Immunity," *Southwestern University Law Review* 12 (1981).

Silver, Roxanne L. and Camille B. Wortman. "Coping with Undesirable Life Events." In *Human Helplessness: Theory and Applications,* edited by Judy Gerber and Martin E. P. Seligman. New York: Academic Press, 1980.

Soble, Stephen M. "A Proposal for the Administrative Compensation of Victims of Toxic Substance Pollution: A Model Act," *Harvard Journal on Legislation* 14 (1977).

Solomon, Norman and Harvey Wasserman. "We All Live Downwind," *Environmental Action* (April, 1983).

Strand, Palma. "The Inapplicability of Traditional Tort Law Analysis to Environmental Risks: The Example of Toxic Waste Pollution Victim Compensation," *Stanford Law Review* 35 (February, 1983).

Thode, E. Wayne. "Tort Analysis: Duty-Risk v. Proximate Cause and the Rational Allocation of Functions between Judge and Jury," *1977 Utah Law Review* no. 1 (1977).

Tilevitz, Orrin E. "Judicial Attitudes Toward Legal and Scientific Proof of Cancer Causation," *Columbia Journal of Environmental Law* 3 (1977).

Titus, A. C. "Government Responsibility for Victims of Atomic Testing: A Chronicle of the Politics of Compensation," *Journal of Health Politics, Policy and Law* 8 (Summer, 1983).

Trauberman, Jeffrey. "Compensating Victims of Toxic Substances Pollution: An Analysis of Existing Federal Statutes," *Harvard Environmental Law Review* 5 (1981).

Treadway, Laura. "When a Vet 'Wants' Uncle Sam," *American University Law Review* 31 (1982).

Truman, Jay. "A Guide to A. C. T. B.," *Downwinders* (1982).

————. "The View from Downwind," *Testing News* (1982).

Ulrich, Helmuth. "The Incidence of Leukemia in Radiologists," *New England Journal of Medicine* (1946).

Wehr, Elizabeth. "Bill to Help Radiation Victims Collect Damages Bogs Down," *Congressional Quarterly* (July 3, 1982).

————. "Victims of Radiation, other Environment-Related Illness Seeking Help from Congress," *Congressional Quarterly* (February 23, 1980).

Wyant, Mary F. "The Discretionary Function Exception to Government Tort Liability," *Marquette Law Review* 61 (1977).

Yalow, Rosalyn S. "Reappraisal of Potential Risks Associated with Low-Level Radiation," *Annals of the New York Academy of Sciences* (November, 1981).

Notes

"The Application of the Statute of Limitations to Actions for Tortuous Radiation Exposure," *Alabama Law Review* 31 (1980).

"Baneberry Widows Want Wrongful Death Case Reopened," *Las Vegas Review* (May 5, 1984).

Citizen's Voice 2, no. 2 (Fall 1983).

"Karen Silkwood, Federalist," *Regulation* 8 (January/February, 1984).

"The Occurrence of Malignancy in Radioactive Persons," *American Journal of Cancer* 15 (1931).

"Tort Actions for Cancer," *Yale Law Journal* 90 (1981).

Reports

Atomic Energy Commission. *Atomic Test Effects in the Nevada Test Site Region* (January, 1955).

————. *Leukemia Mortality Studies in Southwest Utah*, by Edward S. Weiss (1965).

————. *Observed Relations Between the Deposition Level of Fresh Fission Products from Nevada Tests and the Resulting Levels of I-131 in Fresh Milk*, by Harold Knapp (March, 1963).

Banbury Report No. 4. *Cancer Incidence in Defined Populations*, by John Cairns, Joseph L. Lyon, and Mark Skolnick. New York: Cold Spring Harbor Laboratory (1980).

Department of Defense. *Operation Buster-Jangle, 1951, U.S. Atmospheric Nuclear Weapons Test, Personnel Review*. DNA 6-23F (1982).

————. *Operation Ranger, 1951, U.S. Atmospheric Nuclear Weapons Test, Personnel Review*. DNA 6-22F (1982).

————. *Operation Tumbler-Snapper, 1952, U.S. Atmospheric Nuclear Weapons Tests, Personnel Review*. DNA 6-019F (1982).

————. *Operation Upshot-Knothole, 1953, U.S. Atmospheric Nuclear Weapons Test, Personnel Review*. DNA 6-014F (1982).

Department of Energy. Eighth Dose Assessment Advisory Group Meeting (October 20, 1983).

Energy Research and Development Administration. *Final Environmental Impact Statement, Nevada Test Site, Nye County, Nevada.* (September, 1977).

National Conference on Environmental Injury Compensation. *The Superfund Section 301 (e) Study Group Recommendations and Related Pending Proposals*, by George Freeman, Jr., Washington, D.C. (March, 1984).

GAO Report, Vol. 1, EMD-81-1. *Problems in Assessing the Cancer Risks of Low-Level Ionizing Radiation Exposure* (January, 1981).

Institute of Electrical and Electronic Engineers. *Low Dose Radioactivity, Chromosomes and Cancer*, by John Gofman, et al. (October, 1969).

Interagency Task Force on Compensation for Radiation-Related Illnesses. *Draft Report*, by William G. Schaffer, Chairman, Washington, D.C. (1980).

National Council on Radiation Protection, 16th Annual Meeting. *A Review of External Radiation Dose Estimates Offsite Associated with Nuclear Testing at the NTS and Current Research*, by McDonald E. Wrenn (April, 1980).

Radiation Effects Research Foundation. Second United States–Japan Workshop. *Reassessment of Atomic Bomb Radiation Dosimetry in Hiroshima and Nagasaki.* "The Use of Thermoluminescence Analysis for Atomic Bomb Dosimetry: Estimating and Minimizing Total Error," by E. H. Haskell, P. L. Kaipa, and M. E. Wrenn (1983).

Special Task Force for the Governor of Colorado. *Report of Rocky Flats Employees Health Assessment Group.* Denver, Colorado (March 21, 1984).

SECOPA Meeting (1982) *Hard Judicial Choices,* by Philip J. Cooper (November, 1982).

U.S. Congress. House Committee on Interstate and Foreign Commerce, Subcommittee on Oversight and Investigations. *The Forgotten Guinea Pigs: A Report on Health Effects of Low-Level Radiation Sustained as a Result of the Nuclear Weapons Testing Program Conducted by the United States Government.* 96th Congress, 2d session, August, 1980.

————. Joint Committee on Atomic Energy. *Atomic Energy Legislation Through the 94th Congress.* 94th Congress 2d session, 95th Congress 1st session, March, 1977. Appendix (. Index to the Legislative History of the Atomic Energy Act of 1946.

————. *Soviet Atomic Espionage.* 82nd Congress 1st session, April, 1951.

————. Special Committee on Radiation of the Joint Committee on Atomic Energy. *The Nature of Radioactive Fallout and Its Effects on Man.* 85th Congress, 1st session, May 27–29, June 3–7, 1957.

————. *Superfund Study Group, Minority Report* prepared by Judge Charles D. Breitel. 98th Congress, 1983.

Utah Cancer Registry. *Cancer in Utah, 1957-1974,* edited by Charles R. Smart and Joseph L. Lyon. Report No. 1 (September, 1972), Report No. 2 (November, 1975). Salt Lake City, Utah (September, 1979).

————. *Cancer in Utah, 1966-1977,* edited by Charles R. Smart, Joseph L. Lyon, and Harmon J. Eyre. Report No. 3. Salt Lake City, Utah (September, 1979).

Congressional Hearings (listed chronologically)

Senate Special Committee on Atomic Energy. *Atomic Energy,* pt. 1. 79th Congress, November 27–30, 1945.

Joint Committee on Atomic Energy. *Atomic Energy.* 81st Congress, 1st session, February 2, 1949.

————. *Health and Safety Problems and Weather Effects Associated with Atomic Explosions.* 84th Congress, 1st session, April 15, 1955.

————. *The Nature of Radioactive Fallout and Its Effects on Man.* 85th Congress 1st session, May-June, 1957.

————. *Fallout from Nuclear Weapons Tests.* Summary of Hearings. 86th Congress, 1st session, August, 1959.

————. Subcommittee on Research, Development and Investigation. *Radiation Standards, Including Fallout.* 87th Congress, 2d session, June 4-7, 1962.

House Committee on Interstate and Foreign Commerce. Subcommittee on Health and the Environment. *Effects of Radiation on Human Health—Health Effects of Ionizing Radiation.* 95th Congress, 2d session, 1978.

House Committee on Interstate and Foreign Commerce, Subcommittee on Oversight and Investigation, and Senate Scientific Research Subcommittee of the Labor and Human Resources Committee on the Judiciary. *Health Effects of Low-Level Radiation.* Joint Hearings. 96th Congress, 1st session, April 1979.

Senate Committee on Veterans Affairs. *Veterans' Claims for Disabilities from Nuclear Weapons Testing.* 96th Congress, 1st session, June 20, 1979.

House Committee on Interstate and Foreign Commerce. Subcommittee on Oversight and Investigations. *Low-Level Radiation Effects on Health.* 96th Congress, 1st session, April, May, August, 1979.

Senate Committee on the Judiciary and Committee on Labor and Human Resources. Subcommittee on Health and Scientific Research. *Radiation Exposure Compensation Act of 1979*. Joint Hearings. 96th Congress, 2d session, June 10, 1980.

Senate Committee on Labor and Human Resources. *Radiation Exposure Compensation Act of 1981*. 97th Congress, 1st session, October 27, 1981, 97th Congress, 2d session, April 18, 1982.

Newspaper Articles

Denver Post
11 April, 1984

Deseret News (Salt Lake City, Utah)
25,26,29,31 October, 1950
6 November, 1950
27,28,29,30 January, 1951
1–4,6 February, 1951
4 November, 1951
5,17–21,25 April, 1952
8,9 May, 1952
17,18,26,27 March, 1953
29 April, 1953
1,20,21 May, 1953
8 March, 1955
4,13 May,1955
26,27 April, 1957
27,30 May, 1957
25,26 June, 1957
3,22 July, 1957
17 September, 1957
8 October, 1957
17 September, 1958
1,27,29 October, 1958
29 August, 1978
23 October, 1978
13 November, 1978
8,23 December, 1978
30 January, 1979
8 February, 1979
12,13,25 May, 1984

Iron County Record (Cedar City, Utah)
16 October, 1946
1 January, 1948
3 August, 1950
4 January, 1951
1,4,29 November, 1951
5,19 March, 1953
7,21,28 May, 1953

4,25 June, 1953
24 February, 1955
3,10 March, 1955
30 May, 1957
6 June, 1957
8 August, 1957

National Catholic Reporter
23 April, 1982

New York Times
18 July, 1945
26,31 October, 1958
19,31 December, 1958
7 April, 1983
8,9,13 May, 1984

Rocky Mountain News
11 April, 1984

Salt Lake Tribune
21 January, 1955
27 May, 1978
29 August, 1978
7 December, 1978
25 April, 1984
9,11,12,16,18,26 May, 1984.

Washington County News (St. George, Utah)
25 January, 1951
1,8 February, 1951
1 March, 1951
1 November, 1951
3 April, 1952
17 February, 1955
3,24 March 1955
14 April, 1955

Washington Post
8 January, 1979

Court Cases

Irene Allen, et al. v. United States. 527 F Supp 476 (1979, 1981, 1982)

Irene Allen, et al. v. United States. Civil No. C-7a-055-J (May 10, 1984)

Sylvia Barnson, et al. v. Foote Mineral Col, Vandium Corporation, and the United States. Civil
       No. 80-119 (U.S. District Court, Utah, C.D.D.)

Bartholomae v. United States. 253 F2d 716 (1957)

John N. Begay, et al. v. United States. Civil No. 80-982 (U.S. District Court, N.D.)

Bulloch v. United States. 145 F Supp 824 (1956)

————. 95 Fed Rules Dec 123 (1982)

————. 82-2215, 82 2352 (Civil No. C-81, 0123C) (November 23, 1983)

Dalehite v. United States. 346 US 15 (1953)

Doe v. McMillan. 412 US 306 (1973)

Feres v. United States. 340 US 135 (1950)

Home Box Office (HBO) v. Wilkinson. 531 F Supp 986 (1982)

Jackson v. Kelly. 557 F2d 735 (1977)

Krumback v. Dow Chemical Company and Travelers Insurance Company. No. 82 CA 1443
       (September 1, 1983)

Kuhne v. United States. 267 F Supp 649 (1967)

Ozark Airlines v. Delta. 402 F Supp 687 (1975)

Rayonier v. United States. 352 US 315 (1957)

Silkwood v. Kerr-McGee Corporation. 104 SCT 615 (1984)

Smith v. United States. 375 F2d 243 (1967)

United Airlines v. Weiner. 335 F2d 379 (1964)

United States v. Union Trust Company. 221 F2d 62 (1955) Affirmed by U.S. Supreme Court
       350 US 907 (1955)

Index